THE FAMILY, MEDICAL DECISION-MAKING, AND BIOTECHNOLOGY

Philosophy and Medicine

VOLUME 91

The title published in this series are listed at the end of the volume.

THE FAMILY, MEDICAL DECISION-MAKING, AND BIOTECHNOLOGY

CRITICAL REFLECTIONS ON ASIAN MORAL PERSPECTIVES

Editor

SHUI CHUEN LEE

National Central University, Chungli, Taiwan

 Springer

A C.I.P. Catalogue record for this book is available from the Library of Congress.

R
725.5
.F24
2007

ISBN 978-1-4020-5219-4 (HB)
ISBN 978-1-4020-5220-0 (e-book)

Published by Springer,
P.O. Box 17, 3300 AA Dordrecht, The Netherlands.

www.springer.com

Printed on acid-free paper

TABLE OF CONTENTS

ACKNOWLEDGEMENTS

This book grows out from the proceedings of the Third International Conference of Bioethics on *The Ethical, Legal and Social Implications of Human Pluri-potent Stem Cells* held on June 24–28, 2002 and the Fourth International Conferences of Bioethics on *Biotechnology, Family and Community* on June 24–26, 2004 at National Central University and National Taiwan University, Taiwan. They were jointly organized by the Graduate Institute of Philosophy of National Central University and the Medical School of National Taiwan University, the Center for ELSI Research in Taiwan, the Project for the Enhancement in the Humanities and Social Sciences, Advisory Office, Ministry of Education. The conferences are generously supported in funding by the National Science Council, the Health Department, the Ministry of Education, the National Research Program for Genomic Medicine, National Science Council, the Mainland Affairs Council, the Chinese Development Fund, the National Health Research Institutes Forum, and the National Central University Foundation.

The first International Conference of Bioethics was first proposed by the Graduate Institute of Philosophy of National Central University in 1998 and received continued support both from local and overseas institutes. Our first local co-organizer, the Graduate Institute of Philosophy of Nanhua University of Chia-Yi; and overseas institutes: Institut für Asienkunde, Hamburg, Germany; the European Academy for Environment and Economy, Hamburg, Germany; and the Governance in Asia Research Centre (formerly Centre for Comparative Public Management and Social Policy) of Hong Kong, City University of Hong Kong; have been offering generous and continuing support throughout the years. Local and overseas scholars from every part of the world who joined and contributed in these conferences have greatly enriched our exchanges, especially on topics and differences between the East and West on bioethics. The conference logistics are mostly carried out by a distinctive group of graduate students led by Mr. Hon Chung Wong of the Graduate Institute of Philosophy, and their selfless devotion was what made these conferences successful and fruitful as they were. The proceedings were edited again with the help of Mr. Hon Chung Wong and others throughout the years.

Thanks are due to the speakers and authors of the papers of the conferences and many of the revised papers collected here have benefited from the participators of these conferences. Special thanks must go to Professor H.T. Engelhardt, Jr., who is not only one of the keynote speakers of our last four conferences, but is also the benefactor of the editing of this volume. It is his suggestions for the preliminary planning of the book and his endless effort to push the editing process that makes this volume a reality. I am deeply indebted to his never-ending encouragement in both the organization of our conferences and the editing ofs this and other books on bioethics. Many thanks to Professor Ruiping Fan, the series editor of *Asian Studies in Bioethics* and *The Philosophy of Medicine 2*, for his friendship and help both in preparation and in the writing of our Introduction to this volume. My personal thanks also to Mr. Justin Ho, the assistant to Professor H.T. Engelhardt, Jr., of Rice University, in Houston, Texas, for his efforts in the editing of this book and his contribution to the major part of the writing of our Introduction to this volume. Last but not least, I would like to register my gratitude to the two former principals of Central National University, Professors Chao-Han Liu and Chuan-Sheng Liu, for their continuing supports and endorsements to these conferences and other projects of my Institute.

Shui Chuen Lee

CHAPTER 1

MEDICINE AND THE BIOMEDICAL TECHNOLOGIES IN THE CONTEXT OF ASIAN PERSPECTIVES

SHUI CHUEN LEE AND JUSTIN HO

I. A VIEW FROM ASIA: AN INTRODUCTION

The rapid development of biomedical technology in the last two or three decades has been felt in both the west and in the east. Some might even argue that Asian countries have been more impacted by these developments because of the transformation of traditional society brought about by modernization. However, some scholars have claimed that most Asian countries or regions (especially Japan, Korea, Singapore, Taiwan, Hong Kong and mainland China) have shown a somewhat different pattern of response to such biomedical innovations. Such persons claim that Confucianism is the backbone behind the dynamics of such responses. They assert that Asian countries have a strong commitment to family ties while the west has developed a strong liberal individualism. These scholars argue that such familial factors affect and provide different perspective on bioethical issues, from the practice of health care to the development and application of biotechnologies.

This volume provides access in English to contemporary East Asian biomedical reflection. It draws primarily from resources framed by and responding to Confucian moral and philosophical concepts. East Asian medicine, biomedical research, and health care policy are shaped by their own set of moral and cultural commitments. Chief among these is the influence of Confucian ideas. A portrayal is offered of the implications of Confucian moral and ontological understandings for medical decision making, human embryonic stem cell research, and health care financing. The result is a multifaceted insight into what distinguishes East Asian bioethical reflections. The volume opens with an exploration of the Confucian recognition of the family as an entity existing in its own right and which is not reducible to its members or their interests. As the essays in this volume show, this particular appreciation of the family supports a notion of family autonomy that contrasts with certain Western individualistic accounts of proper medical decision making. There are analyses of basic concepts as well as explorations of their implications for

1

S.C. Lee (ed.), The Family, Medical Decision-Making, and Biotechnology, 1–13.
© 2007 *Springer.*

actual medical practice. The conflicts in East Asian countries between traditional Confucian and Western bioethics are explored as well as the tension between the new reproductive technologies and traditional understandings of the family. East Asian reflections concerning the moral status of human embryos and the morality of human embryo stem cell research disclose a set of concerns quite different from those anchored in Christian and Muslim cultural perspectives. The volume closes with an exploration of how Confucian cultural resources can help meet the contemporary challenges of health care financing.

II. THE FAMILY AND ITS IMPACT ON BIOETHICS AND PERSONHOOD

The volume first introduces the role the family plays in Confucian approaches to medical decision making. Confucianism regards the family as having an ontological independence such that the authority and welfare of the family cannot be reduced to the authority and welfare of its constituting members. This Confucian appreciation of the family underlies the family-oriented character of medical decision making in much of East Asia. The articles in the section also contrast the Confucian view of the family's role in medical decision making with the views currently prominent in North American and European bioethics.

Ruiping Fan in "Confucian Familism and Its Bioethical Implications" provides a portrait of Confucian familism and its application to bioethics. According to Confucianism, the family is not only a means for human flourishing; it also has its own ontological reality. One's relatives sustain an element of one's own being and the family is one of the essential constituents and structures of the universe. This entails that one's choices have a profound effect on both the family as a whole and on the world. The world is impacted in that everyone according to Confucianism has "a seed of virtue" or what might be thought of as the potential to become virtuous. When family members cultivate this seed of virtue by acting in accordance with the duties and rituals associated with their roles in the family, the whole world is made virtuous.

This deep metaphysical understanding of the family is reflected in the Confucian approach to making medical decisions, assessing medical technology, and financing health care. Medical decision-making involves the whole family, since the whole family's interests as well as the individual's are at stake. The interests of the family must also be taken into account when assessing medical technology. This does not mean that individual interests should always give way to family interests when conflicts occur. Confucianism advocates appealing to virtue to mediate for virtue in conjunction with the duties associated with familial roles require that families be primarily responsible for their members' health care needs.

This Confucian view of the family is, of course, just one view of the family and its role in the good life. As H. Tristram Engelhardt, Jr., points out in his article, "The Family in Transition and Authority: The Impact of Biotechnology," currently there is considerable dispute about the nature of the family and the role that the family plays in human flourishing. According to Engelhardt, traditional views of the family, like the ones exemplified by Confucianism, tend to be rooted in metaphysical commitments. The family and its members are seen from a cosmic perspective so that the family is also regarded as an essential part of reality necessary for human flourishing. Post-traditionalists, however, "tend to regard the universe and humans as ultimately coming from nowhere, ultimately going nowhere, and for no ultimate purpose." Having rejected such rich understandings of human history, post-traditionalists "tend to regard each other as moral and metaphysical strangers." Autonomy, individualism, and free choice are regarded as valuable and are often pursued to the detriment of the family structure. In fact, the family is viewed by some post-traditionalists as an impediment to realizing the interests of its members. As a result, attempts to maintain and restore traditional views of the family have led to conflicts between attempts to maintain versus reform traditional views. As Engelhardt notes, the view of the family to which one subscribes has a strong impact on health care policy. For example, the growth of liberal individualism in America and the West has given rise to policies promoting isolated, individual, self-directed choice. However, this has the consequence of reducing the meaning of the family to its members, allowing for the restructuring of the family through consent and the use of new reproductive technology to pursue individual fulfillment. As this volume shows, it is important to examine the role of medical technology in promoting liberal individualism and the questions it raises regarding the role of the family in human flourishing.

The impact of secular liberal individualism, globalization, and biotechnology on the family and the self is also the focus of Stephen Erickson's article, "Family Life, Bioethics, and Confucianism." Stephen Erickson, who holds a traditional view of the family, assumes that all aspects of human nature are not contained in the world. That is, he holds the view that human beings belong to and derive their existence from something other than this world. If biotechnology evolves to the point where every aspect of human nature is producible in this world, then that transcendental aspect will forever be lost. Erickson also believes that this transcendental aspect of human life proceeds from our life on earth, and that the family is integral to help us achieve our full status as human beings. However, liberal individualism, globalism, and biotechnology threaten this transcendental dimension of human life by disrupting the traditional structure of the family.

Nevertheless, Erickson argues that a philosophical view such as Confucianism has the resources necessary to retain the transcendent aspect of human existence that is so central to our identity. Unlike liberal individualism, Confucianism views the family as central to personal identity and flourishing. It is through *hsin*, sentiment, or empathy that we experience such notions as "self" and "person", with the family serving as the principal mediator of this dimension of human life. An essential dimension of whom and what we are is the result of how we are experienced and treated by our family. Therefore, nourishing *hsin* may help to prevent the loss of the self in the face of the growing threat of liberal individualism, biotechnology, and globalism.

III. MEDICAL DECISION-MAKING AND TRADITIONAL CONCEPTION
OF THE FAMILY

The implications of the Confucian view of the family for actual medical decision making are wide-ranging. In many Asian countries that have been influenced by Confucianism, the family continues to play a central role in medical decision-making, so that, in many cases, physicians first disclose the patient's diagnosis to the family and it is then up to the family to decide whether to disclose this information to the patient. The first two articles in this section explore whether this practice is ethical and whether and what attempts should be made to reform this practice.

In the "Moral Ground of Truth Telling Guideline Development," Shuh-Jen Sheu examines what factors should be taken into consideration when constructing a guideline for determining whether, what, and how much information doctors should disclose to terminally ill cancer patients in Asian countries like Taiwan. As Sheu notes, in many Asian countries like Taiwan, doctors still hesitate to tell the truth to patients even though studies have shown that most patients want to know the truth. Sheu acknowledges that the practice of lying to patients has a negative affect on the patient's ability to make autonomous choices. However, it is also important to understand the reasons families choose to lie. Families want to prevent the patient from feeling despair and hopelessness and the patient is often believed to be incompetent or vulnerable. But more importantly, in countries like Taiwan, individual autonomy is not regarded as central to personal identity. In addition, the family often feels morally obligated to be protective of ill members. Consequently, any guidelines for telling the truth to terminally ill patients should also take into account such things as cultural perspectives and require health care professionals to be aware of the cultural background of patients and understand why patients make the decisions they do. A better understanding of the patient's feelings and concerns when trying to communicate effectively with the patient will help policy-makers and doctors to make better decisions about when autonomy should give way to paternalism and vice versa.

In his article, "Truth Telling to the Sick and Dying in a Traditional Chinese Culture," Stephen Wear uses a consequentialist approach to argue against the view that in cultures like China, which emphasize the family as a decision-making unit, lying to patients should be an ethically guiding principle. Wear draws on Sissela Bok's argument in *Lying: Moral Choice in Public and Private Life*, to illustrate how the harms that might arise from lying to patients can outweigh any benefits. Bok acknowledges that some have argued that it is difficult to inform patients properly so they can make good medical decisions, that telling the truth can distress patients, and that depriving patients of hope may lead them to withdraw from treatment and hasten death, and that lying to patients can allow doctors to make appropriate decisions, respect the role of families, and choose more compassionate courses of action. Bok, however, argues that patients are capable of understanding information if it is properly presented and that the notion that patients will die prematurely lacks empirical support. Bok also argues that not knowing the truth can be torturous for patients, and that patients often come to know the truth anyway as their condition deteriorates. Further, the family and physicians may make choices that collide with the patient's values. Deception may also make it more difficult to provide appropriate aggressive treatment to patients. Telling the truth on the other hand might allow patients to exercise choices that are in accordance with their values, enhance compliance and coordination, avoid needless pain and suffering, and improve the quality of treatment.

Wear, however, does not insist that truth-telling become the rule in countries like China, arguing that doing so (1) might not be well received and (2) might undermine the traditional Chinese family. Current studies show that in China, the family has not been undermined as the appropriate decision-maker, so that the presence of the family may improve the success of medical care. Undermining the family might deprive patients of the support and succor provided by families. Given these considerations and given the fact that patients are a diverse group, which includes people who favor and ask for a traditional familist approach, and those who do not but fail to voice their preferences, Wear advocates asking patients what sort of truth-telling approach they favor. This elevates the patient to the status of "key person" in the arena of medical decision-making. If asked, physicians should advocate the familist approach, since it would allow families to provide patients with support and avoids alienating them at a time when their active presence may prove beneficial to the patient.

In the next article in this section, "On Relational Autonomy: From Feminist Critique to Confucian Model for Clinical Practice," Shui Chuen Lee argues that we should abandon individualist theories of autonomy in favor of what he calls an "ethical relational theory of autonomy." Lee notes that medical decision making policies are often rooted in individualistic theories of autonomy. However, feminists claim to have shown that all individualistic theories of autonomy are in some way problematic. Procedural theories of autonomy, which claim that an agent is autonomous "if she has subjected her motivations and action to the appropriate kind of reflection," cannot distinguish certain non-autonomous actions from autonomous actions. Weak substantive theories of autonomy, which claim that agents are autonomous if their beliefs and desires have certain content, do not specify what the substantive content is. Strong substantive theories require that agents be competent and able to distinguish from right and wrong. And while these are an improvement over other theories of autonomy, they fail to realize that one's identity cannot be understood without taking into account one's relations with others. Consequently, Lee holds that we ought to reconstruct strong substantive individualistic theories into what he calls an ethical relational theory of autonomy.

The "ethical relational theory of autonomy" integrates the Confucian concept of a person, which asserts that our relations with others, and in particular our family, are part of our personal identity. This theory of autonomy also has an ethical component: it takes into account the Confucian insight into the nature of moral experience, which as Lee shows, is quite similar to the Kantian notion of autonomy. Lee argues that an autonomous action is an action that (1) is circumscribed by the "moral mind" or what would Kant terms "practical reason" and (2) this moral mind must be oriented to the welfare of others because their wellbeing is closely linked to our own wellbeing and identity. This second feature of this theory takes into account the family, because our moral practice begins with the wellbeing of family members as they are so integral to our own identity and wellbeing. Lee concludes by arguing such a theory requires that medical decision making be a collective affair that involves both patients and their families.

In "Regulating Sex Selection in a Patriarchal Society," Wenmay Rei begins with the circumstance that biotechnology has evolved to the point where parents can choose the sex of their child. Questions, however, arise as to whether parents should be allowed to select the sex of a child in strongly patriarchal, Confucian cultures like those found in Taiwan. While sex selection to prevent hereditary sex-related disease is typically not seen as being morally problematic, this is not the case when we consider certain sex selections done for non-medical reasons. For example, sex selection done for sexist reasons is seen by many as morally objectionable. This concern has led Taiwan to publicly forbid sex selection based on non-medical reasons. However, as Rei points out it is not clear that sex selections performed for non-medical reasons such as protecting one's child from sexism or because of a purely independent preference for a child of a particular sex are not morally justifiable unless we can show that serious negative consequences will ensue if the practice is forbidden. Nevertheless, because it is difficult to ascertain the motives of parents, one might still think a formal ban is warranted.

Rei, however, argues that in the case of Taiwan, because sexism is deeply embedded in its informal structures, unless there is a democratic consensus, a formal ban is not likely to be effective. Until social attitudes change, women will still face a great deal of social pressure to have a son and will simply turn to abortion or having more children if sex selection for non-medical reasons is formally banned. Moreover, Rei claims that enforcing such a ban might also have a negative effective on women's procreative freedom and autonomy.

Finally, in "Modern Technology and Reproductive Technology" Leonardo D. de Castro identifies and explores two types of challenges to the Confucian conception of the family by developments in modern medicine. The first type of challenge has to do with extending the frontiers of human freedom while the second has to do with extending the frontiers of human responsibility. De Castro focuses on advances in reproductive technology to illustrate the first type of challenge and advances in organ transplantation to illustrate the second kind of challenge. Reproductive technology opens up a plethora of new options that bring with them considerable benefits. However, along with these benefits arise certain consequences; one result is that the family becomes redefined and certain values become redefined. The question then arises; should human freedom be curtailed in such cases? Much of the paper is spent illustrating how advances in organ transplantation generate new responsibilities. Technology has progressed to the point where organ transplants involving organs from unrelated persons are highly successful. Given this fact and given the shortage of organs and the inability and unwillingness of many family members to contribute, it seems as though the duty to supply organs to those who need them has extended beyond the realm of one's genetic family.

IV. EMBRYONIC STEM CELL RESEARCH: CONFUCIAN, ISLAMIC, AND WESTERN PERSPECTIVES

As biotechnology continues to advance, ethical questions arise concerning whether and when particular technologies may morally be engaged and what limits should be placed on their use. The articles in this section are concerned with whether stem cell research is morally acceptable. Embryonic stem cell research (ESC) theoretically carries much promise. Embryonic or pluripotent stem cells are capable of developing into any type of body tissue and have the potential to be used to repair damaged organ tissues. Many maladies such as spinal cord injuries and Parkinson's disease may be curable with the use of such cells, though as of yet there is little concrete success. However, because stem cell research using material from embryos always involves the destruction of the embryo, the decision as to whether allow ESC and the methods that should be used to derive the stem cells necessary for ESC are the subject of much controversy.

Many scholars have observed that discussions in East Asia regarding research with human embryonic stem cells have a character quite different from that found in North America and Western Europe. The essays in this section claim that the roots of this difference lie in the contrast between approaches grounded in Confucian moral and philosophical concepts versus those grounded in the moral-theological commitments of Abrahamic religions.

In "The Ethics of the Embryonic Stem Cell Research and the Interests of the Family," Ruiping Fan argues that the liberal strategy of offering an account of individual rights so as to place the individual in authority to make moral decisions fails to provide a persuasive argument for ESC research because it is a version of ethical individualism. Fan distinguishes between weak liberal individualism and strong liberal individualism and then draws out their implications for stem cell research. According to the weak version of individualism, the individual has sole authority in the decision-making process but individual self-determination is not intrinsically valuable and so society has no obligation to promote self-determination. Strong individualism holds that exercising self determination is intrinsically valuable for the individual and respecting individuals requires strengthening their capacity for exercising self-determination. Fan then argues that neither version of ethical individualism places any set of moral constraints on how adults choose to produce and treat human embryos. Adherence to the weak versions implies that since embryos lack the capacity for self-determination and society has no obligation to promote self-determination, we are entitled to experiment on embryos so long as we are able to secure the consent of those who created the embryos. Although the life of the embryo is necessary to becoming an individual who can exercise self determination and the strong version of ethical individualism holds that self determination is intrinsically valuable and should be promoted, it does not follow that the life of

the embryo is dominantly intrinsically valuable. Fan argues that because the life of
the embryo is not a sufficient condition for self determination, it must be balanced
against other values. Therefore, the strong version also holds that it is the consent of
individuals, which ultimately determines what should be done to human embryos.
Confucians, according to Fan, would view the weak version of ethical individualism
as incomplete because while it holds that self-determination has so high a status that
society may not interfere with it, it does not hold that self-determination is a value
that ought to be promoted. Confucians would also claim that the strong version is
incoherent because self-determination is seen as intrinsically valuable yet this view
does not entail that there should be any constraints placed on the production and
use of embryos.

Fan then argues that Confucian ethical familism offers a two dimensional
ethical strategy, which takes into consideration both individual interests as well as
family interests and offers a persuasive argument for ESC research that escapes
the problems associated with liberal ethical individualism as it places reasonable
constraints on the use and production of human embryos for research. Fan holds
that this ethical strategy suggests that it is permissible to conduct stem cell research
on the embryos left over from IVF treatments because the family has the authority
to decide how many children they would like to have and because these embryos
are going to die in a short time anyway and much good may come from their use. It
is also permissible to conduct therapeutic cloning to secure embryonic stem cells to
save individual family members because they are more important than the embryo in
terms of what they have contributed to the family's integrity and survival. However,
Fan notes that ESC research whose intention is to benefit some individuals and
families in the future may not be justified according to Confucianism because
(1) it may not be precisely righteous (2) there is no urgent family interest at stake and
(3) the gain achievable from this research is much less certain than the gain that
could be obtained from therapeutic cloning when it is already technologically
reliable.

In "A Confucian Evaluation of Embryonic Stem Cell Research and the Moral
Status of Human Embryos," Shui Chuen Lee examines whether ESC research is
morally permissible from a Confucian Perspective. Lee argues that what is central
to this issue is the moral status of the embryo. Lee first examines Mary Anne
Warren's account of the moral status of the embryo, which claims that because
early stage embryos lack sentience and are not moral agents, they only have the
same moral status as other non-sentient beings. This suggests that it is morally
permissible to engage in ESC research. Lee, however, claims that while this account
is compelling, it still suffers from some problems. It fails to take into account those
who hold that it is wrong to experiment on embryos because these entities have the
same moral status as fully developed humans. According to Lee, the interests of
these individuals matters because they are also members of the moral community
since Confucianism claims that those for whom one feels sympathy are members
of the moral community. Since the interests of the individuals who are concerned
for embryos matter, we must treat embryos with respect and use embryos only

when there are no other means to procure stem cells. This suggests that the status of embryos is in some sense elevated when we take into account the concept of moral community. However, Confucianism sees no problem with using embryos left over from IVF treatment, since these were created for procreation and will ultimately be destroyed anyway. While their destruction is in some ways still a tragedy, their destruction will at least be to the benefit of others. Further, while Lee acknowledges that we must not use therapeutic cloning to produce embryos if there are other methods to acquire stem cells in order to respect those who regard human embryos as having the same moral status as fully developed human beings, therapeutic cloning may still be justified in cases where such cloning can lead to cures for significant diseases. In such cases, the continual pain and suffering of families provide a compelling reason to use cloned embryos.

In his article, "Regulations for Human Embryonic Stem Cell Research in East Asian Countries: A Confucian Critique," Hon Chung Wong notes that policies regarding the derivation of stem cells are more liberal in East Asia as opposed to the United States and most European countries. In China, for example, the derivation of stem cells from cloned embryos and research on combining human somatic cell nuclei with animals are permitted. However, in some European countries, no human embryonic stem cell research is permitted or research is only permitted on embryos no longer needed for reproduction. Wong attributes this to the fact that many Western countries view embryos as having the same status as persons and consequently, think that they have the same rights as persons. Wong examines these various policies from a philosophical perspective. He is critical of the commonly held view of personhood because he believes in order to be a person one must have memory and self-identity. Early stage embryos, which have not developed a nervous system, lack both of these features. His criticism extends to countries which place a time limit on when available embryos from IVF treatment can be used; if embryos that result from these treatments are going to be destroyed anyway, then such a time limit, according to his account, would not be warranted if the use of embryos can contribute to the well-being of future generations. He then moves on to examine the policies on stem cell research in East Asian Countries. His view is similar to Lee's in that he holds that because we all live in a moral community, we have strong moral commitments to each other; therefore, we need take the interests of those who object to stem cell research into account. Further, we should not create tragedies unless we have good reason to do so. Consequently, like Lee he favors using therapeutic cloning to derive stem cells from embryos only when no other means are available and only under certain circumstances.

The concluding two papers underscore the differences in perspective between Confucian understandings and those lying at the roots of Islamic and Christian cultures. In "Stem Cell Research: An Islamic Perspective," Sahin Aksoy, Adurrahman Elmali, and Anwar Nasim examine ethical issues related to embryonic stem cell research, drawing on Islamic insights. According to these authors, Islam is not only a religion but a way of life "which is based on the recognition of the unity of the Creator and our submission to His Will." Whether to allow for ESC research and which methods for deriving ESC are morally permissible, ultimately hinges on the moral status of an embryo, which, in turn, can be ascertained by appealing to the Qu'ran and Hadiths, sayings of the prophet Muhammad whose primary role are to help us better understand the verses of the Qu'ran. According to the Qu'ran and two hadiths, the body and soul combine to form a "full human person." However, this does not occur at conception but around forty to fifty days after conception. For this reason, terminating the life of an embryo before ensoulment occurs "is regarded as disliked (makruh), while it is considered forbidden (haram) after this stage."

This has several implications for stem cell research. For one thing, there is no problem using adult stem cells. In the case of fetal stem cells that are derived from spontaneously aborted fetuses or leftover embryos from IVF treatments, it is also permissible to derive stem cells from these sources as they would be disposed of anyway. And in the case of IVF embryos, they have not been ensouled yet. However, it is unlawful to induce abortions for the purpose of deriving stems cells if the fetus is younger than fifty days, and it is forbidden to induce abortions for the purpose of deriving stem cells if the fetus is older than fifty days. It also should be noted that Islam subscribes to the notion that human life has great value and should not typically be treated as a means to some end. Therefore, it is impermissible to create embryos solely for the purpose of stem cell research and every effort should be made not to create spare embryos for IVF treatments to avoid wasting human life.

As Engelhardt shows, traditional Christianity, in contrast to Confucianism, strongly forbids the destruction of the early human embryo. It is this Christian moral insight that provides the cultural source of the special character of the stem cell debates in the West. Christians of the first thousand years and Orthodox Christians today recognize morality to be grounded in a noetic experience of God, which shows the taking of human life, even at early embryonic stages, to have a moral significance as serious as that of murder, whether or not the embryo is ensouled. As Engelhardt indicates, this position is in accord with Orthodox Jewish Halakha regarding the obligations of Gentiles. Jewish views in the matter of abortion are often poorly understood, in that there is a failure to distinguish between what Jews hold to be morally acceptable to Jews, and that which is morally acceptable for Gentiles, sons of Noah. The difference lies in the circumstance that Gentiles are bound by the law given to Noah and his sons, unlike Jews, who are bound by the law given to Moses. As a result, from the prohibition against shedding blood, included among the seven precepts given to the sons of Noah (*Sanhedrin* 56[ab]), the Halakha for Gentiles (ben-Noah) forbid abortion as a wrong punishable by execution. "On the authority of R. Ishmael it was said: [He is executed] even for the murder of an embryo. What is R. Ishmael's reason? –Because it is written, *Whoso sheddeth the blood of man within [another] man, shall his blood be shed*. What is a man within another man? – An embryo in his mother's womb" (*Sanhedrin* 57[b]). Christianity recognizes that the law as applied to Gentiles now clearly applies to all. This moral appreciation is grounded in a noetic theological epistemology tied to the recognition that the universe has ultimate meaning in being grounded in an experience and presence of a personal Creator God. This moral and epistemological framework is at radical contrast with that of Confucian thought and helps to frame the difference between debates regarding the morality of human embryonic stem cell research in the Pacific Rim versus that in North America and Western Europe.

V. TAKING THE FAMILY SERIOUSLY: CONFUCIAN APPROACHES TO HEALTH CARE

Every country is faced with the task of adopting a health care system that is not only economically sound but also just. The papers in this section deal exclusively with the issue of health care financing from a Confucian perspective. In her paper, "Confucian Health Care System in Singapore: A Family-oriented Approach to Financial Sustainability," Kris Su Hui Teo compares the health care systems in Singapore and Hong Kong, arguing that Singapore's health care system is both economically sound and consistent with Confucian ethics. Hong Kong currently has a welfare-based health care system, which provides highly subsidized health care services to its citizens. Teo claims that because costs are rising, taxes are low, and more people are getting older, this system may soon become unsustainable. Further, the quality of services at public facilities is poor, the time spent waiting for treatment is lengthy, and people often have no choice but to accept what services are offered. While the government has recently created the Mandatory Provident Fund (MPF), a pension fund, this measure fails to remedy these problems as the amount one contributes is small in comparison to the cost of health care.

Singapore, on the other hand, has a three-tiered health care system. Employees and their employers contribute money to the Central Provident Fund (CPF), a social security program. Some of the money from the CPF is then placed in medical savings accounts, which can be used to pay for health care services. Second, some of the money is also used to buy catastrophic medical insurance to help pay for injuries or illnesses that exceed the amount in one's medical savings account, and to provide for those who are over the age of forty and can no longer work due to disability or old age. Third, the government provides health care subsidies as a last resort for those who cannot afford to pay for medical care. Not only is Singapore's health care system fiscally sound but because families purchase health care services themselves, they have more treatment options and can purchase high quality services if they so desire. Also the government is able to divert more funds to improve the quality of health care offered by public facilities.

Teo also argues that such a system is consistent with Confucian ethics in making the family the primary source of health care. Confucianism, a virtue based ethical theory, asserts that we must strive toward *Ren* or benevolence. We strive toward *Ren* by fulfilling our obligations to those whom we have ties with. According to Confucianism, one must first extend *Ren* to one's closest family members before gradually extending it outward to non family members. Extending *Ren* to one's family members is called *Xiao* or filial piety. Singapore's health care system is in line with *Xiao* and *Ren* since it allows the money from medical savings accounts to be transferred to one's immediate family members. The fact that employers must help their employees to cover their medical costs by offering insurance in Singapore is also in line with *Ren*. Rulers have obligations to aid their subjects according to Confucianism, and they exercise *Ren* when they fulfill those duties. The relationship between employers and employees is consistent with such a relationship. Finally, the fact that government steps in to help those people who cannot afford health care as a last resort is line with *Ren* as this is a condition of a benevolent government.

Erika H. Y. Yu, in her paper, "Respect for the Elderly and Family Responsibility: A Confucian Response to the Old Age Allowance Policy in Hong Kong," argues that the current old age security system in Hong Kong is problematic because it is not fiscally sound or consistent with the Confucian virtue of *Xiao*. Yu limits her criticism to the Old Age Allowance Policy. As she notes, the number of elderly persons who receive Old Age Allowance is steadily growing. Given the low tax rate, it is likely that the current welfare scheme will not be financially sustainable, and over time the quality of service will likely diminish. Further, this program also lacks moral vision. The Old Age Allowance was installed to achieve the following objectives: (1) to provide partial assistance for the elderly, (2) to encourage families to care for elderly members, and (3) to reduce the burden of the elderly on their families. Yu argues that the second and third objectives stand in opposition of the Confucian virtue of *Xiao* or filial piety. Because the Old Age Allowance may allow family members to shift part of the family's duty to care for elders to public resources, the Old Age Allowance does not, in fact, encourage families to take care of elderly. Rather the Old Age Allowance might lead families to view society as the primary caretakers of the elderly. The Old Age Allowance might also reinforce the notion that being elderly is a burden since children are obligated to support elderly people who are strangers to them. This is significant because for Confucians, *Xiao* demands that (1) children and not the government take care of their parents when they become old, and (2) children should not view caring for their parents as a burden. This leads Yu to suggest the abolishment of the Old Age Allowance program. To help elderly people whose families cannot afford their care and to help those families cultivate *Xiao*, Yu proposes that the government give tax breaks to such families and facilitate employment. However, if financial subsidies are needed, they should be means-tested by taking the children of the elderly as a unit for assessment.

Justin Ho in his essay entitled "Is Singapore's Health Care System Congruent with Confucianism?" examines whether Singapore's system is, in fact, as consistent with Confucianism as Teo claims. Specifically, his essay examines whether Singapore's approach to health care is contrary to the Confucian view of an ideal society. Ho notes that by making the individual and his or her family the primary bearers of choice and responsibility in health care financing and decision-making, Singapore's healthcare system does create space for the central Confucian values of *ren* (a) and *xiao*. However, Ho points out that Singapore's health care system also provides government-subsidized health care to the medically indigent, (i.e., those persons who do not have the financial resources to pay for healthcare), which may be inconsistent with the Confucian conception of an ideal society. According to Confucianism, an ideal society is one where the community – not the government – bears primary responsibility for providing healthcare to persons. Furthermore, Ho observes one might also argue that the mandatory character of medical savings accounts is deeply contrary with the Confucian ethical and political ideal of a limited government. Ho, however, notes that even if Singapore's health care system does in fact conflict with this ideal, if such a system serves a higher cause, such as the harmony of the state and universe, then such measures might nevertheless be justified as the end of all human action according to Confucianism is to bring about the harmony of the universe as a whole and the harmony of the state is a necessary condition for the realization of this end.

VI. CONCLUSION

Few persons would argue against the view that families generally endow their individual members with the trust and support that can only arise from close and intimate relationships. However, as we have seen, many of the authors in this volume hold a stronger view, namely that familism should play a key role in shaping our medical institutions and practices. If this latter view is correct, further theoretical implications follow. For instance, as we have seen, recognizing that the participation of family members is appropriate implies that the family ought to be included in the medical decision making process and that the family should consent to genetic information collection and research. Moreover, familism has the consequence of transforming the notion of individual autonomy and self-determination into a notion of ethical relational autonomy, which in turn gives further justification to the legal rights of family participation in social and in an individual's affairs. Some of the adherents of familism would also argue that the family introduces a third sector into the health care policy that mitigates some of the dilemmas between a public coverage and the individualistic market provision of health care needs.

Before concluding, it should be noted that some of the scholars in this volume such as Shui Chuen Lee also claim that taking methodological familism seriously does not entail losing sight of the benefits of individualism. They argue that in the social and political field, we maintain the rights and responsibilities of the individual as a moral agent and when internal conflict occurs among family members, Confucians never take the interest of the family as overriding. When familism produces internal conflicts, we ought to fall back on the principle of *ren* and evaluate the situation at hand against our moral conscience or against the suffering faced by each member of the family. Such scholars note that Confucius and Mencius depicted a number of instances where a filially pious son need not blindly follow the order of the father. Therefore, in cases of insoluble family conflicts, the family may be regarded as temporarily dissolved and in these situations the patient has the last word and should not be unduly pressured by others.

CHAPTER 2

CONFUCIAN FAMILISM
AND ITS BIOETHICAL IMPLICATIONS

RUIPING FAN

City University of Hong Kong, Hong Kong

I. INTRODUCTION: WHY CONFUCIAN BIOETHICS IS SO DIFFERENT

The taken-for-granted character of traditional Confucian bioethics is radically at odds with that of the dominant modern North American and Western European models. The latter regards individuals as ideally atomic, as persons who should determine themselves autonomously, that is, in their own terms and absent the manipulation or suasion of others. In some fashion, Western individuals are to come to the self-possession of their own view of right conduct and human flour-ishing, outside of history, culture, and their place in society. As a result, physicians are taught to help patients choose without the influence of family members. For instance, spouses and children are asked to leave and then the physician asks the patient whether the patient wants to have all discussions in absence of and held in confidentiality from all others. The contemporary, dominant, secular view of bioethics takes as its normative and theoretically guiding point of departure the autonomous, free, and self-responsible individual.

The difficulties with this account are numerous. To begin with, no such person exists. Persons always exercise their choices within the socio-historical context, thickly influenced by particular ideologies, power structures, and moral under-standings.[1] Confucian moral and metaphysical thought acknowledges all of this and more. It recognizes that humans, or at least most humans, grow up within families. It recognizes that all humans carry with them the taken-for-granted assumptions of their socio-historical circumstances. The claim is more robust. Confucian moral thought regards persons in general and patients in particular as located within families because Confucian thought appreciates the deeply family-centered character of social reality. It is not merely that persons flourish within families. More impor-tantly, families have a social and ontological reality of their own, so that their members are never adequately understood, nor their good ever adequately appre-ciated or for that matter realized, save within the family. What the West no longer

S.C. Lee (ed.), The Family, Medical Decision-Making, and Biotechnology, 15–26.
© 2007 *Springer.*

recognizes is central to the truth appreciated by Confucian tradition, namely, that the family is central to social reality.

Thus Confucian bioethics is in fundamental tension with the liberal, social-democratic, as well as individualistic bioethics. A sufficient understanding of the differences requires an acknowledgement of the extent to which Confucian moral and bioethical commitments are nested in a thick set of metaphysical and axiological commitments often underappreciated in the West. This article lays out the geography of these commitments. It does so by focusing on what is termed familism in order to craft a contrast with individualism. First, unlike the moral individualism of the West, Confucian moral and metaphysical thought appreciates that families, not individuals, have ontological priority. It is not just that individuals only flourish within families. It is rather that families are the grounding reality within which individuals must be understood. The fundamental symbolism of this Confucian familism is found in the *Classic of Change* (Section II). Moreover, there is an ultimate power that sustains the Confucian familism, which power is virtue, not entitlement. Confucian thought presupposes a synergy between human free choice and the power of virtue (Section III). The centrality of the family expresses itself in the role played by families in medical decision-making. It expresses itself as well in the various ways in which patients are at least initially shielded from the unpleasant character of their diagnosis, so as better to lodge their treatment within the knowledge that only a family can have of its members (Section IV). The reality of the family is not only acknowledged at the micro-level, but it is enshrined at the macro-level in Confucian-influenced health care policy, such as one finds in Singapore (Section V). Finally, Confucianism appeals to a complex, two-dimensional Confucian moral strategy (taking into account both individual and family interests) for assessing biotechnologies. This strategy contrasts with liberal, individualistic, moral strategies (Section VI).

II. THE FEATURES OF CONFUCIAN FAMILISM

Confucian familism is grounded in the recognition of the resonance between the reality of the family and the deep reality of the universe. Classically, this is appreciated in terms of not merely foundational concepts, but grounding symbols and sustaining rituals. The cardinal symbols, it must be stressed, disclose in a quasi-mystical fashion the deep character of reality. That is, there is a deep metaphysical congeniality between the character of the symbols and the character of reality. This core Confucian metaphysical and epistemological thesis is illustrated by the fundamental symbolism of Confucian familism, disclosed in one of the Confucian classics compiled by ancient Confucian sages, the *Classic of Change*. This classic uses 8 trigrams, 64 hexagrams, and 384 lines as symbolic representations in order to display the essential constituents and structures of the universe, both nature and society.[2] Among them is the family: the symbolism shows that proper family relations and activities embody a way of life that carries a cardinal human virtue – the power of shared life. Specifically, the 37th hexagram, *jia ren* (family members), indicates the basic Confucian symbols of the family: a stove (fire) used to cook food for a family to live together in a house as well as virtue (wind) emanated from, manifested in, and spread out of this shared life.[3]

Classical Confucian understanding of family life is not merely determined by any particular economic or social development. Nor can a proper familist way of life automatically occur without the teaching and nurturing of the sages. It is understood that there was a time in which people were well fed, warmly clad, and comfortably lodged, but they lived improperly – they lived almost like beasts. It was the sages who managed to lead people to live a true human life in appropriate family relations (*Mencius* 3A4: 8). In the 37th hexagram, the text (the interpretations of the hexagram as well as its each line), and in the *Commentaries (Xiang Zhuan and Twan Zhuan)* on it, we find the normal features of the family as confirmed by Confucianism.[4]

First, the husband and the wife have their correct places in the family, and this is the great righteousness shown in the relation and positions of heaven and earth (*Twan Zhuan*). Specifically, the husband must regulate the family with sincerity, honesty, and majesty, setting a good example for other family members to follow (lines 5 & 6; *Xiang Zhuan*). Moreover, what is most advantageous is that the wife be firm and correct (General Interpretation), attending to family affairs (such as preparing food) carefully (line 2). In short, the husband-father and wife-mother together rule the family, like an authoritative ruler for a state (*Twan Zhuan*).

Second, family regulations should not be loosened if the family wants to avoid occasion for repentance (line 1). Stern severity may sometimes hurt members, but there will usually be good fortune; on the other hand, if the wife and children are smirking and chattering all the time, shame and difficulty will ultimately be brought to the family (line 3).

Finally, "let the father be indeed father, and the son be son; let the elder brother be indeed elder brother, and the younger brother younger brother; let the husband be indeed husband, and the wife be wife – then will the family be in its normal state" (*Twan Zhuan*). In its normal state, the family can be enriched by every member, especially by the wife-mother, and there will be great fortune (line 4). If the family is thus established and enriched, all that is under heaven will be established and enriched (*Twan Zhuan*).

Indeed, the Confucian takes these family activities as the essential elements of political life. Even if one does not work in government, one is still taken as being engaged in government as long as one is filial to one's parents and takes care of one's brothers (*Analects* 2:21). If families are well established, the whole world can be made peaceful (*Great learning*). In addition to the husband-father, the wife-mother is crucially important in the family because she actually manages all important family affairs, including educating the children. She can especially enrich the family because she is in the position of distributing the resources that the husband-father gains from labor outside the family. The promise of the family depends on the moral integrity, wisdom and self-restraint that she possesses in bringing all family members together in a warm and pleasant family life.

How is this familist way of life possible? Confucianism recognizes that, although humans naturally live together like many other social animals, they may not live together properly. Among other things, perverse sexual intercourse has been common in human societies. Humans must follow a series of systematic behavioral patterns in order to become civil. Confucians have found the paradigms of such patterns in the primary rituals or rites (li) that were established by the sages: for a man and a woman to have sexual intercourse properly, a marriage ritual must first be performed; for a 20-year-old boy to become an established responsible person in the family, a capping ritual must be performed; for a passed-away parent or grandparent to be treated properly, the family must conduct a burial ritual; and for treating the family's deceased parents, grandparents, and ancestors properly, the family must offer sacrificial rituals. These rituals must be performed with serious commitment and attention. It requires that one attain an attitude of reverence in performing the rituals (*Analects* 2:7; 3:26). Confucius could not bear to see "the forms of mourning conducted without real grief" (*Analects* 3:26). In short, ritual behavior incarnates the grounding Confucian symbols of the family. The rituals are not only sustained by the symbols. They also sustain the symbols. The congeniality between symbol and deep reality is progressively disclosed and progressively appreciated through ritual behavior, supporting the connection between humans and deep reality.

III. THE POWER OF VIRTUE

Sustaining symbols and observing rituals require virtue. Confucians recognize that virtue (de) is a rightly-directed energy. It is not simply a habit or a set of dispositions that aid in the pursuit of good and right conduct. It aims one in pursuing the good and realizing the appropriate full being of a human. Virtue is not only embodied in a particular way of life, but it is also generated from an encounter with the good and the noble. Hence, virtue is not an entitlement assigned to everyone by government, but is internal disposition, quality, character or power. Indeed, for Confucians, virtue is first and foremost a power that enables one to control passions, observe rituals, and preserve the family. Great virtue can give one psychic energy, moral force and charismatic influence over others.[5]

Where is such virtue from? Mencius argues that the seed of virtue has been implanted into everyone's heart/mind by Heaven, the divine, so that everyone is already partly a virtuous person (*Mencius* 2A4:3; 2A4:5). Confucius may have agreed to this point when he states that human nature is similar while practices are different (*Analects* 17:2). But the seed must be cultivated in order to become the true fruit of virtue. For Confucius, the way of observing the rituals is the way of cultivating virtue (*Analects* 12:1). Here it seems that a circular argument has been engaged: in order to observe the rituals, one needs virtue; in order to acquire virtue, one must observe the rituals. Circular reasoning can be broken by recognizing that everyone, with the seed of virtue, can choose to cultivate virtue by learning and observing rituals.[6] The better one learns and performs the rituals, the more virtue one acquires. If one chooses not to observe the rituals, one's seed of virtue is wasted: one cannot become truly virtuous.[7] It is for this reason that Confucian gentlemen (junzi) must instruct people to promote virtue and preserve their families by learning and observing the rituals.

Confucian thought presupposes a synergy between human free choice and the power of virtue. Humans could not freely turn to virtue, were there not already present in them a recognition of virtue and a disposition to pursue it. Yet, this disposition is not sufficient. Persons must choose on their own to pursue virtue. This choice itself would be insufficient to bring humans successfully into a life of virtue. Their success lies in the energy of virtue, which transforms them when they turn from their passions and self-interest to pursue that which is truly human. Moreover, that which is truly human is grounded in the deep character of reality, which itself calls forth virtue. The Confucian ritual account of virtue has a deep metaphysical commitment to recognizing that humans find themselves in a relationship with the power of Heaven, which power responds to a rightly-oriented affirmation of virtue.[8] This relationship, which is at its roots embodied in symbol and ritual, will progressively nurture and promote virtue. Individuals do not only pursue the good and avoid the violation of right-making conditions, but achieve a profound harmony with the nature of deep reality.

The nurturing of virtue and, in particular, the nurturing of humans through the power of virtue itself in reality is achieved through ritual. Confucius teaches that we should recognize that every turn to virtue in life is a result of participating rightly in ritual, so as to orient ourselves to the deep character of reality. Role-filling ritual performances transform humans and the reality within which humans find themselves. "Behave in public business as though you were conducting an important guest-accepting ritual. Deal with the people as though you were officiating at an important sacrificial ritual" (*Analects* 12:2). In this way, one's virtue will be bright.[9] That is, one will by engaging rightly in ritual gear into the deep core of reality itself.

In short, the Confucian familist way of life has its strength because it is in harmony with the deep character of reality and therefore the power of virtue that preserves it. The family is by no means merely a reality in a particular time and place. It has continuity over time and in fact exists as a sacred entity transcending generations. That is, individuals come and go, but the family persists. The rituals in which the family engages bring the family members together from the past through the present into the future. In the present, while remembering the past and offering sacrifices to the dead, the family members of this present turn to the family's future. As a result, all the family members are collected into the unity of one focus: the integrity, survival, and prosperity of the family. Confucians see the eternal value of individuals as embedded in this prior reality of families, which for its part is preserved by divine power itself.

IV. SHARED FAMILY-DETERMINATION FOR HEALTH CARE

The rich and thick strength of the Confucian understanding of the family is illustrated in its model of shared family-determination for issues important to every family member. This includes even today a shared responsibility for health care decision-making. So, too, the education of children, career choices, and the selection of marriage partners are not so much the choice of individuals as an expression of the wisdom and concern of the family as a whole. Confucian reflection recognizes the family both in authority to make such decisions and as the best authority to make such choices. Its exercise of authority expresses the prior moral and ontological reality of the family. Its exercise of authority reflects the developed insights and maturity of the family's judgment. This point of view lodges medical decision-making not in the isolated choices of the patient, but in the decisions of the family as a whole. A family representative bears the weight of acting as intermediary between the physician and the patient (Cong, 2004; Fan, 1997; Fan, 2004).

This authority and obligation extends even to at first protecting the patient from a truth, and then aiding the patient to come to terms with the illness and the needed treatment. If the family judges that it is important pro tempore or for the long term to shield the patient from the trauma of an ominous diagnosis, the family appropriately requires the physician to protect the patient from the truth until the family can bring the patient to collaborate appropriately in the needed response to the illness (Fan, 2002; Fan & Li, 2004). The family is to work as a whole in the process of health care decision-making for any ill family member.

It is not easy to affirm and maintain this Confucian understanding in the face of Western bioethics that has even produced Westernized individually-oriented understandings of Confucian familist practice. Liberal individualism is increasingly pressing all cultures to conform to its terms. It is attempting to establish itself as the politically correct ideology. Arrogant Western liberal scholars seek to export to Confucian societies their understandings of individual autonomy, independence, and the subjectivity of values. This is done under the cover of advancing so-called universal human rights and global norms of bioethical conduct. Others criticize the Confucian familist model of medical decision-making as ethically backward and un-enlightened in rhetorically loaded terms. Still others attempt through consequen-tialist analogies to demonstrate the benefits of frank disclosures and unsophisticated truth-telling, as prevails in North America and Western Europe. They place all this within an account of the history of medical ethics in which medical paternalism is successfully replaced by a robust self-determination. In contrast, they judge Confucian society as not just lagging behind the course of history and failing to catch up with advanced Western counterparts, but as more significantly embedded in a wrongly-directed false consciousness. Arrogant Western bioethicists and moral theorists often select unfavorable cases from Confucian society to demonstrate the "obvious" defects and shortcomings of the familist model: "one can easily imagine that she might have made a different [much better] choice than the one her family made for her."

The appreciation of the truth of Confucian moral and bioethical understanding is not grounded simply in a consequentialists assessment of costs and benefits. It is rather achieved in an experience grounded in an attitude of reverence that opens the heart, or as the Greeks would say, the nous, to the reality of Confucian familism.

This reality is for Confucians a deeply metaphysical truth. The husband and the wife, as with the fundamental elements of *yin and yang*, are bound together, neither sufficient in itself but both completed in bond with the other. No true humanity is realized outside this unity of yin and yang. It is not simply that man is insufficient without woman, or woman insufficient without man. In becoming husband and wife and forming a family, they enter into a reality in which their existence as man and woman is radically realized as husband and wife.

Similarly, the roles of parent and child are to each other complementary. Only in the union of husband and wife bound in the procreation of the child is the family fully achieved. Outside of this relationship, humans are one-sided and incomplete. In this sense, the family is internal to every normal adult individual. In contrast, with liberal individualism – as Steven Erickson succinctly indicates – individuals are points without substantive relationship to others. Although they may form human relationships, such relationships do not become internal to the complete adult. They are experienced, understood, and encountered more as external clothing rather than as internal flesh and bone (Erickson, 2007). In Confucian society, close relatives are not merely blood kin or socially supportive; they sustain an ontological element of one's very being. As a consequence, one cannot take account of their opinions en passant in making one's individual choices. In order not to be one-sided and incomplete, one's decisions must be made together with them. In special health care situations, families make decisions for patients without the participation of the patients.

When individualists criticize Confucian familism by selecting particular unwise family decisions made on behalf of a patient in order to criticize Confucianism, they ignore or discount the numerous cases of unwise individual decisions supported by and endorsed within the secular individualistic West. Indeed, when commenting on decisions within their own cultural and moral ambit, they often proceed on the view that individuals are in authority over themselves and rightly should decide concerning themselves, even if their decisions harm themselves and the members of their family. That is, they regard individualistic choice as reflecting a right-making condition, which gives authority and moral status to the decisions independently of the consequences. However, they do not consider the possibility that family-embedded medical decision-making may itself reflect a right-making condition, giving moral worth to family-oriented decisions independently of the consequences for the individual patients. In short, they operate with a double standard.

The point is that the Confucian account is grounded in a radically different sense of human flourishing and of the moral life. It understands not simply that caring, wise, and self-sacrificing families will often choose wisely, but that submitting to a familist approach to decision-making will nurture a virtue that will transform the character of those who embrace it. In addition, because the family is recognized as having an intrinsic moral status, to act in a familist fashion is to act in accord with a fundamental, right-making condition. Beyond that condition of right action, Confucians recognize positive consequences from familist decision-making. Many Chinese love their families not simply because in the long run this turns out to be advantageous, but rather because of this love they enter into a dimension of human flourishing and virtue otherwise undisclosed. This is not to deny that there can be abuses and that special precautions and protecting rules should not be established. For instance, to protect its moral integrity and to recognize its own medical limits, a medical institution in a Confucian society may be obliged insist that, if a family's decision is egregiously in discord with the physician's professional judgment about the medical best interests of the patient, the physician should communicate directly with the patient (Fan & Li, 2004). Though one may put such special protective rules in place, nevertheless most patients can be left free from the harm of an imposed "right" to know the truth since they live within a functioning family and enjoy the virtues it secures.

V. FAMILY RESPONSIBILITY FOR HEALTH CARE FINANCING

Contemporary Western health care policy is both financially and morally bankrupt. It is financially bankrupt in the straightforward sense that it has created an extensive web of entitlements to care that outstrip the available financial resources. With a larger proportion of the population beyond retirement age, there are now insufficient workers to generate the funds as transfer payments to those who are no longer working and in need of social support and medical treatment. The financial challenges are made even more difficult by the increasing number of very elderly (i.e., over 80), requiring more long-term nursing as well as high-cost medical interventions. All of this is morally complicated by a twin loss of a sense of obligation to save as a family to meet these financial challenges and to act as a family whole to support members through various challenges and periods of life.

The result has been the establishment of a liberal, social-democratic ethos, with political and moral commitments that have led to a focus on egalitarian outcomes and the creation of health care entitlement that in the long run cannot be sustained. Faced with limited resources and driven by a totalizing egalitarianism, either such cultures must level down health care options, as in Canada, or they must affirm a range of entitlements that cannot be financed, as in Western Europe and the United States. This state of affairs is intimately connected to media democracies that are exposed to the political hazard of politicians seeking election by offering entitlements that in the future will not be financially sustainable (Engelhardt, 2006). Rather than establishing entitlements or affirming an egalitarian ethos, the Confucian moral and political understanding supports a pursuit of virtue through creating financing schemes that emphasize family security and family choice. Such an approach is illustrated by the Singaporean approach to medical savings accounts (Teo, 2007).

The focus on family-oriented medical savings accounts, the resources for which can be used across the family and which can be bequeathed to future generations, follows the Confucian maxim: to enrich people is to enrich families. This approach requires specific roles for both family and government. At the level of the family, Confucianism assigns importance to the proper role of the wife-mother in managing family affairs. At the level of government, Confucianism insists that rulers should not tax people more than ten percent of their income so that most resources can be left to families for pursuing their own flourishing. Indeed, the Confucian viewpoint for public policy and governance is radically familist: populate, enrich the people, and educate them (*Analects* 13:9). When a governor complains that he cannot follow the Confucian rule of only tithing the people, because he does not have enough resources even when he taxes them with two-tenths, a disciple of Confucius replies: "if the people do not have plenty, with whom can the governor have plenty?" (*Analects* 12:9).

Confucians have always argued that a good government must tax or levy lightly (e.g., *Mencius* 1A5:3). It is not that Confucians do not understand that tithing the people cannot afford enough resources for the state to afford equal welfare and health care for everyone. It is rather that they don't think it is right for the state to do so. When Confucians appreciate the family life as the basic way of human existence, they take it for granted that families bear primary responsibility for their members' welfare and health care. The Confucian virtue implies that the parents must take care of their children, and adult children must take care of their elderly parents.[10] In this regard, the Singaporean family-savings-based health care system sets a heuristic example: Confucian familism is not only morally ideal to pursue, but it is also practically feasible and realizable for a fruitful health care delivery in contemporary society (Teo, 2006). In this respect, taxation policies in Singapore and Hong Kong succeed better than in most developed countries in leaving disposable income in the hands of families.[11]

VI. BIOTECHNOLOGY AND THE FAMILY

Medical science and technology should be developed insofar as this strengthens the family. Those medical technologies that enable the family successfully to reproduce therefore, all else being equal, will be affirmed. Confucian population policy supports having children, not limiting procreation. The Confucian family reproductive ethos is in stark contrast with the dominant reproductive ethos of liberal societies, where the allure of double-income, childless families moves husbands and wives to pursue their own self-indulgence rather than the procreation of children and nurturing them. As the challenges to health care policy in Western Europe and North America indicate, the failure to support families is not only morally corrupt but can also be economically disastrous. However, most of the difficulties with reproduction cannot be solved by better technology, but by a more virtuous understanding of the nature and mission of the family. Moreover, all technologies must be morally assessed in terms of their impact on the family.

It is for this reason that such new biotechnologies such as human cloning conflict with the Confucian moral perspective, because they set at jeopardy the traditional understanding of the family, in which husband and wife join to produce the children of the family. A man and a woman joined as husband and wife are thus

foundationally normative to understanding all sexuality, and most especially reproductive sexuality. As Confucians see it:

> Heaven and earth existing, all things then got their existence. All things having
> existence, there came male and female. From the existence of male and female there
> came husband and wife. From husband and wife there came father and son. From father
> and son there came ruler and minister. From ruler and minister, there came high and
> low. When the distinction of high and low had existence, there came the arrangements
> of ritual and righteousness.[12]

At stake is not simply natural genealogy, but a Confucian appreciation of what is
socially and morally normative. First, the existence of male and female reflects the
profound normative relation of Heaven and earth, which may never be changed.
A human is either a male or a female. Even if there are deficient and ambiguous
cases (e.g., persons of aberrant genotype or sexual phenotype), such cases must be
appreciated as falling short of realizing that which is normatively human. Instead,
the normative, complementary ontology of being male and female allows the
complementary interaction of husband and wife, which lies at the foundation of the
possibility of the family. Moreover, family generations carry eternal moral meaning.
Parent is parent, child is child, and a child must be from both father and mother.
These distinctions are never to be erased. If reproductive human cloning blurs these
distinctions in society, Confucians must support its prohibition in society.

Because of the normative character of the family founded in the union of husband
and wife, biotechnology that would undermine them as the reproductive source of
families would set at jeopardy a basic source of human flourishing: the family.
It is for this reason that Confucian thought is opposed not only to reproductive
cloning, but to any form of genetic engineering that would alter the character of
human sexuality and the normative relationship of husband and wife. The same can
be said of concerns regarding destructive research with human embryos, including
the acquisition of human embryonic stem cells. The ethos of such interventions
may subtly bring into question the normative character of Confucian concerns for
children, so that all things being equal, there should be a burden of proof against
such research and the use of adult stem cells should be favored instead. Confucian
familism will favor assessing such technologies not in terms of a simplistic utilitarian
calculus, but instead in terms of a concern for maintaining a particular ethos, a
particular sense of relationship within the family and to one's children.

A Confucian appreciation of new reproductive technology requires first and
foremost, considering the interests of the family together with the interests of the
individual. It does not endorse sacrificing individual interests to family interests as
a general moral solution. Instead, the pursuit of virtue must guide in the preservation not just of the interests of the family, but of the ethos of familism itself.
An ideal world would not have conflicts between individual interests and family
interests. But our non-ideal world and our often corrupt society generate conflicts
between individuals and families. The Confucian approach must be to avoid such
conflicts when possible and mediate such conflicts when they are unavoidable.
Where Confucian thought might accept certain technological interventions (such

as human embryonic stem cell research), such interventions must nevertheless be recognized as not ideal (although justifiable and permissible in some non-ideal situations), because they potentially threaten the Confucian familist ethos. This complex, two-dimensional Confucian moral strategy for assessing biotechnologies contrasts with liberal, individualistic, moral strategies (Fan, 2007). It realizes that choices are often not simply either right or wrong; there are categories of choice that are understandable and tolerable, even though they fall short of the full marks of virtue.

VII. CONCLUDING REMARKS

The character, content, and texture of Confucian bioethics are at loggerheads with those of the individualistic agendas born of the American bioethics of the 1970s. Unlike the secular individualist moral reflections of the West, Confucian thought appreciates in quasi-religious terms the bond between phenomenal world and noumenal reality. It locates families as enduring realities. In addition, it recognizes that virtue is gained through a ritually structured family life. The acquisition of virtue is thus not simply the straightforward result of instruction. Instead, it is acquired within a familist way of life that gears one into the very fabric of deep reality and is accomplished through ritual behavior, which places one in harmony with deep reality itself.

ACKNOWLEDGEMENTS

The research on "Confucianism and Parent-Child Relation" covered in this essay was funded by the Governance in Asia Research Centre of City University of Hong Kong, Hong Kong in 2005.

NOTES

[1] Even the bioethics that emerged at the beginning of the 1970s in North America bears the unmistakable imprint of the social upheavals of the time. For various reasons, individuals were attempting to find what they held to be their true selves outside of the authority and influence of traditional social structures. The spirit of that time was counter-cultural and post-traditional. So, too, is the bioethics it produced. But all persons, including such bioethicists and their bioethics, are embedded in particular socio-historical contexts.

[2] Tradition has ascribed the 8 trigrams to legendary King Fuxi (c. 3000 B.C.), the 64 hexagrams to King Wen (c. 1171-1122 B.C.), the text to Duke Zhou (d. 1094 B.C.), and the Commentaries to Confucius' disciples and their disciples (500–300 B.C.). For a brief English introduction to this classic, see Chan (1973, pp. 262–263).

[3] (Li lower, Xun upper; Jia Ren). The stove is used to prepare food that is shared by all family members. Food "sharing" is much more literally true in Confucian families than in other types of families: every dish is shared by everyone. Many visitors of China are first impressed and then get used to this way of Chinese eating. This is just one typical indication of the sharing of Confucian family life. Finally, through sacrificial rituals, food is also shared with the family's ancestors.

[4] The moral interpretations of the hexagram offered in this essay have followed Cheng Yi (1033–1107) and Zhu Xi (1130–1200) with their influential commentary works on the Classic of Change: *Yi Cheng Zhuan* and *Zhou Yi Ben Yi*. All my English citations to Confucian classics in this essay are adapted form James Legge's translations.

[5] "One who rules with virtue is like the polestar, which remains in its place while all the lesser stars do homage to it" (*Analects* 2:1). "The virtue of the superior is wind. The virtue of the people is grass. Let the wind be over the grass and it must bend" (*Analects* 12:19). "If you desire what is good, the people will at once be good" (*Analects* 12:19).

[6] "Is virtue a thing remote? I wish to be virtuous, and virtue is at hand" (*Analects* 7:19). "Is anyone able for one day to apply his strength to virtue? I have not seen the case in which his strength would be insufficient" (*Analects* 4:6).

[7] Confucius' student Zai Yu did not want to learn and slept during the day time. The Master commented: "rotten wood cannot be carved; a wall of dirty earth will not receive the trowel. This Yu! – what is the use of my reproving him?" (*Analects* 5:10).

[8] Originally de (virtue) is acquired from Heaven by performing a sacrificial ritual to one's ancestors. See Nivison (1994). This is to say, virtue is both given and rewarded by divine powers: it is given as seed endowed in everyone's heart/mind; it is also rewarded when one performs the rituals well. The deep meaning of the 37th hexagram might be this: fire is the divine power that gives and rewards virtue.

[9] The family rituals are still taken to be most important in Confucian cultural societies. In our research on "Confucianism and Parent-Child Relation" conducted in the Spring 2005, we asked primary school Grade Six students in Singapore and Hong Kong about whether the family rituals are important for them to observe, 127 out of 170 students (i.e., 75%) from a Hong Kong primary school and 231 out of 264 (i.e., 88%) from a Singapore primary school answered yes.

[10] This understanding is still at home in contemporary Chinese societies. In our "Confucianism and parent-child relation" survey, 261 out of 264 Singaporean students (i.e., 99%) and 168 out of 170 Hong Kong students (also about 99%) answered that they should take care of their parents when they grow up.

[11] Currently the highest level of income tax is 20% in Hong Kong and 21% in Singapore.

[12] This is from the Section II of a Commentary of the Classic of Change, Xugua Zhuan. See *I Ching, (1973*, pp. 435–436).

REFERENCES

Chan, W.T. (1973). *A Source Book in Chinese Philosophy*, Princeton University Press, Princeton.

Confucius (1971). *Confucian Analects, the Great Learning & the Doctrine of the Mean*, James Legge (trans.), Dover Publications, Inc., New York.

Cong, Y. (2004). *Journal of Medicine and Philosophy*, 29: 2.

Engelhardt, H.T., Jr. (2006). 'Why the United States have had Difficulties Recognizing the Importance of its Private health Care Sector,' in J. Tao (Ed.), *Confucianism and Health Care Market*, Springer.

Erickson, S. (2007). 'Family life, bioethics, and Confucianism,' in S.C. Lee (Ed.), *The Family, Medical Decision - Making, and Bio technology*, Springer.

Fan, R. (1997). 'Self-Determination vs. Family-Determination: Two Incommensurable Principles of Autonomy,' in *Bioethics* (UK: Blackwell Publishers; externally referred international journal), Vol. 11, No. 3&4, 309–322.

Fan, R. (2002). 'Reconsidering Surrogate Decision-Making: Aristotelianism and Confucianism on Ideal Human Relations,' in *Philosophy East & West*, 52: 3, 346–372.

Fan, R. (2007). 'The Ethics of Experimenting with Human Pluripotent Stem Cells and the Interests of the Family,' in S.C. Lee (Ed.), *The Family, Medical Decision - Making, and Bio Technology*, Springer.

Fan, R. and Li, B. (2004). 'Truth Telling in Medicine: The Confucian View,' *Journal of Medicine and Philosophy*, 29: 2, 179–193.

I Ching (1973). J. Legge (trans.), Causeway Books, New York.

Mencius (1970). *The Works of Mencius*, J. Legge (trans.), Dover Publications, Inc., New York.

Nivison, D. (1996). ' "Virtue" in bone and bronze,' *The Ways of Confucianism* (pp. 17–30), Open Court, Chicago.

Teo, K. (2007). 'Confucian Health Care System in Singapore: A Family-oriented Approach to Financial Sustainability.' in S.C. Lee (Ed.), *The Family, Medical Decision - Making, and Bio Technology*, Springer.

CHAPTER 3

THE FAMILY IN TRANSITION AND IN AUTHORITY

The Impact of Biotechnology

H. TRISTRAM ENGELHARDT, JR.

Rice University, Houston, Texas, United States

I. INTRODUCTION: IN THE FACE OF ROBUST MORAL PLURALISM AND CONFLICTING METAPHYSICAL ACCOUNTS

If there is anything that characterizes the contemporary understanding of the family, it is the lack of a common understanding. We share no consensus. Indeed, we live in the face of profound and at times angry disagreements. It is not just that we face a plurality of conflicting moral and metaphysical accounts that had once been geographically dispersed or artificially suppressed. Moral communities with radically different views of the family and human flourishing overlap in the same geographical areas, and their disagreements have become salient. In addition, many once coherent traditional accounts have fragmented, if not fallen into incoherence. We confront not just the post-modernity of numerous competing accounts. We also face a post-modernity born of a loss of focus within once dominating accounts. For many, there is a sense of the center failing, so that it is unclear how to understand the family and its place in the good human life.

For many people for whom traditional accounts are intact, or at least robust in their remembrance, there is a profound sense of a need to respond to the pluralism and incoherence. There is a disposition to a cultural counter-revolution through extending or at least recovering the vigor of traditional accounts often rooted in religious commitments with their deep metaphysical accounts of reality.[1] Against this struggle on behalf of traditionalists of various stripes to protect, extend, or at least restore traditional understandings of the family and its place in human flourishing, there is an equal passion born of the Enlightenment and the progressivist movements of the 19th and the 20th centuries to criticize, revise, and reform traditional understandings in the name of various liberal, post-traditional construes

S.C. Lee (ed.), The Family, Medical Decision-Making, and Biotechnology, 27–45.
© 2007 *Springer.*

of human flourishing that accent free choice, autonomy, and the value of liberty.[2] These accounts and their protagonists seek to recast, if not displace, the traditional family as autonomous and in authority over its members, often displaying a patriarchal structure.

To make matters more complex, there is no single, critical, post-traditional, liberal account of emancipation or liberation from the anti-liberal constraints of traditional accounts. There is not one liberalism, one feminism, or one view of the proper human emancipation from illiberal structures. Despite the passionate, indeed often desperate invocation of consensus, there is dissensus, disagreement, conflict, and controversy. There is a dogged, indeed robust moral pluralism that vindicates Agrippa, the skeptic of the 3rd century A.D., who in his five tropoi argued for the impossibility of resolving such disputes on the basis that (1) philosophical disputes had over the previous 800 years (i.e., from the perspective of the 3rd century A.D.) failed to bring closure to sound rational argument; (2) attempts at sound rational arguments are always nested or lodged within a particular perspective, so that protagonists of particular views argue past each other; all attempts to resolve such debates through sound rational argument require at the outset conceding particular moral premises and rules of evidence so that those in controversy (3) beg the question, (4) argue in a circle, or (5) engage an infinite regress.[3]

All of this is to say that it appears that controversies among irresolvable moral and metaphysical views define the human condition. They always have. Post-modernity, or our contemporary condition however one might want to characterize it, is one of persistent disagreement in the face of the absence of a common way to resolve cardinal disputes by sound rational argument. This observation is not an affirmation of a moral and metaphysical epistemological skepticism, nor a skepticism regarding the existence of truth. It does not involve a foundational metaphysical skepticism. It recognizes, however, the inability of rational reflective discourse to resolve many core controversies regarding right conduct and the nature of human flourishing. We are confronted with important moral and metaphysical disputes that may require special knowledge of an access to the truth for their resolution.

These deep moral and metaphysical disputes are the stuff of the culture wars.[4] The difficulties we have in resolving our important disagreements shed light on the character of our condition, as well as on the context and nature of our discourse. We are confronted with the task of understanding the status, meaning, and authority of the family in the context of health care and with respect to the new reproductive and genetic technologies, all within a circumstance of moral and metaphysical disagreement. We find ourselves at best able to provide a geography of disagreement. This is not a mean or shallow goal. At least like Socrates, we can understand better what we are unable to know as a society compassing numerous and contesting communities, as well as isolated individuals who forward conflicting and incompatible understandings of the family and its place in medicine, all seen within larger concerns with the human good and human flourishing. We will need to note the place of various communities and partisans of particular ideologies in shaping the nature of the disputes. One will need to attend to how the disputes involve levels of revolution and counter-revolution. Traditional Confucian, Christian, Jewish, and Moslem concerns with the family, since they are embedded in a deep metaphysics, tend to have a character distinct in many ways from the controversies engendered by the disagreements among liberal cosmopolitans, libertarian cosmopolitans, feminists of various stripes, and diverse post-traditional offspring of various traditional accounts. After all, traditional accounts tend to place individuals, families, health care decisions, and human flourishing within a cosmic account of human history, human destiny, and human flourishing. Those embedded within traditional accounts tend to have an ultimate sense of where things are coming from, where they are going to, and why. Post-traditionalists tend to regard the universe and humans as ultimately coming from nowhere, ultimately going nowhere, and for no ultimate purpose. The differences between these two genres of perspectives are stark. They separate each side within paradigms, life-worlds, thought-styles, that are deeply incommensurable. There are surely inter-mediate positions, though the extreme poles disclose the depth of the differences at stake.

Consider, for example, the circumstance that traditional accounts tend to invoke accounts of how the family, its meaning, its authority, its significance, and its status are rooted in nature, God, or the deep character of reality, so that there is a recognition of immutable constraints on family structures and their role in human flourishing that lie beyond human choice and agreement. In contrast, despoiled of any such deep rooting in reality, and generally also lacking a rooting in a thick communal understanding of history and human flourishing, post-traditionalists tend to confront each other in the human project as moral and metaphysical strangers. Their accounts of the relationship between individuals, the family, human flour-ishing, and social reality, insofar as they hinge on human choice, human autonomy, human liberty, tend to social arrangements marked by individualism. The accent is on self-realization, self-satisfaction, and individual flourishing, to the detriment of enduring interpersonal and family structures. In contrast, traditional accounts often nurture thickly communitarian views that can recognize a sense of family flour-ishing irreducible to the flourishing of particular individuals. In such accounts, the family becomes a robust intermediate structure between individuals and the state.

II. THE FAMILY AS A MAJOR BATTLEGROUND IN THE CULTURE WARS

The family as a normative and authoritative unit is at best a puzzle and at worst a scandal. It is a puzzle because the family, considered as a fundamental reproductive unit, is a biological given, which new genetic and reproductive technologies can reshape. The successful raising of children usually involves a female with an inseminating male who also often contributes to feeding and protecting the children. This is at best a predominant pattern, and apparently at most a sociobiological fact of the matter. In this context, the question arises: How could anything of normative significance directly follow from a biological given, a usual pattern of behavior? At the beginning of the 21st century within the emerging, global, secular, liberal-cosmopolitan culture, the family, insofar as it reflects biologically grounded behaviors, is viewed as an expression of various merely biological circumstances, which appear as the products of the blind forces of mutation, selective pressure, genetic drift, the constraints of the laws of chemistry, and random cosmic events. Insofar as family structures reflect traditional social commitments, these too are often regarded as arbitrary products of history, if not morally ill-directed. The family as a biological and social unit may be judged to convey both advantages and disadvantages. In particular, within the emerging dominant individualist culture, as well as within social democratic communitarian accounts, this biological given stands as a stumbling block to important goals, such as liberty and equality. As an expression of the forces of biology and history, the family in various ways lies at the roots of patriarchal social structures and conservative moral commitments, which are at loggerheads with the foundational egalitarian and liberty-oriented commitments of the collage of emerging, global, secular viewpoints.

The dominant secular culture also tends to be nominalist with respect to social structures. It recognizes persons, their dignity, their claims to liberty, and their rights to equal treatment as the grounding social and moral realities. This culture can take into account the agreements, contracts, and various social arrangements to which persons give assent. In this context, it is difficult to recognize a biological and historical given as a source or intimation of normativity, much less as a foundation for a social and political structure that might have independent autonomy and authority, as well as patriarchal disposition. A liberally oriented ethos instead regards persons as sources of authority and as cardinal bearers of dignity. Persons are regarded as generating moral claims nested within life projects that may but need not include marriage and the development of families. Such undertakings, as well as the social structures they engender, are interpreted from the perspective of either individuals, or from a general rational authorizing perspective, a particular normative account of persons in community, which is accepted as the source of the value of, and the cardinal place for, the interpretation of such intermediate structures as the family.[5]

Defenders of a liberal social-cosmopolitan ethos recognize the family neither as a neutral biological and social fact, nor as an intermediate social structure with a reality and a standing of its own. As already noted, the family is seen as a stumbling-block to the realization of important moral, social, and democratic interests because it threatens goals and values at odds with individual liberty, as well as with social equality. In particular, the logic of the family is not framed by social justice. Families instead tend to pursue the advantage of their own members. It is not simply that family members have economic and social interests that set families and their members off against the rest of society. Nor is it simply that these interests are likely supported by biologically based psychological proclivities that bind family members together with strong emotional ties inclining them to advantage each other, even if to the detriment of society. In addition, the traditional family[6] is by its very moral logic anti-egalitarian: it is a stumbling-block to fair equality of opportunity.[7]

Notions of the family are multi-valent, and at least five different senses of the family must be distinguished. The first, as already noted, is a cluster of biological or reproductive senses of the family. The family is a basic sexual and reproductive unit associated with the successful rearing of children who themselves can rear children. Generally, this biological unit is composed of a man and one or more women.[8] Its contours may have some sociobiological grounding. As with all biological phenomena, these would involve an interaction of genetics with the environment, which in the case of humans is robustly sociohistorical. Second, over against this biological understanding of the family, there are normative, traditional Christian understandings that serve as a background reference point for Europe and the Americas, and through the dominance of Western Europe and North America for world culture.[9] In their completeness, such normative accounts identify a husband who has headship, who with his wife and minor children constitutes a nuclear family, though this account in its original and its contemporary traditional presentation firmly recognizes a husband and wife without children.[10] The family as a social unit is affirmed not primarily as a reproductive unit but as a basic unit of sexual and social companionship. Third, there are various traditionally grounded non-Christian accounts of the family. Here one might think of the diverse Confucian and other Asian understandings.[11] Fourth, there is a cluster of understandings of family units created by positive law, bestowing on various sexual and reproductive associations legal advantages ranging from welfare support to inheritance rights. Last but not least, there are senses of the family insofar as these can be grounded in liberal cosmopolitan understandings of persons as sources of moral authority and the reference point for goods and interests. This last approach (1) brings into question any normative significance drawn from any particular biological or traditional account of the family as a sexual/reproductive unit, so as (2) to criticize traditional Christian, Confucian, Jewish, Islamic, and other accounts, all in the service of (3) reforming established legal accounts, insofar as these do not derive from the consent of their participants.

This last moral sense or perspective gains social plausibility from technologies that separate sexual activities, reproduction, and child-rearing. These activities, which are in usual circumstances closely associated, as a consequence of the new reproductive technologies appear ever more adventitiously related and as primarily dependent on human choice. A separation between sexual activity and reproduction became dramatically obvious in the mid-20th century with the advent of reliable contraception, along with easily available and safe abortion.[12] The contemporary reproductive technologies enable third parties to be the suppliers not only of sperm but also of ova. Finally, the suppliers of ova, given *in vitro* fertilization and other third-party reproductive technologies and techniques, need not be the person who gestates the zygote produced. Humans now have the opportunity to engage in their reproductive goals with an expanded range of third parties. Cloning goes even further by offering the prospect of an individual's reproducing without the need of a second person's genetic resources. Women may be able to reproduce without a genetic or gestational partner. Men may soon be able to reproduce without drawing on the genetic resources of a woman. The new reproductive technologies substitute human choice for biological necessity. Dependence on biological givens, as well as the standing of traditional views of the family, seems set aside through the technological by enhanced individual choice. As these separations became manifest at the end of the 20th century, traditional construals of marriage and the family were at the same time brought into question as an undertaking of heterosexual individuals, as homosexuals demanded to be accepted at law as the appropriate rearers of children and as partners in a concept of marriage now radically recast.[13] Numerous competing moral and social narratives vie for dominance, creating a plurality of competing accounts of the family.

All this has occurred with considerable controversy. These dramatic revisions in canonical understandings of the family occasion and intensify some of the most significant battles of the culture wars: disputes regarding how appropriately to understand proper human action and normative human social structures. Technology has both freed humans to engage in new choices, as well as opened up strident disputes regarding the kinds of human relationships integral to right conduct and human flourishing. The ability of humans through third-party-assisted reproduction to recast the constellation of expected human sexual, reproductive, and gestational patterns has brought into question the nature of the family itself. These two issues, the one moral, regarding the appropriate ways in which one ought to engage in sexual activities, reproduction, and family life, and the other socio-ontological, regarding the nature of the family, though closely associated, must be distinguished. They intertwine in any appreciation of the family's role for health care policy.

III. THE STANDARD CONTEMPORARY ACCOUNT: INDIVIDUALS AS THE SOURCE OF DECISIONAL AUTHORITY AND THE STRUCTURE OF THE FAMILY

The family is not merely a sexual/reproductive unit. The family is also a social category recognized by many as having independent autonomy and authority, as well as supplying content important for human flourishing.[14] This categorical understanding of the family as constituting a non-reducible dimension of moral reality[15] contrasts starkly with those accounts that reduce the meaning, purpose, and goods of the family to its members, their concerns, and their choices. The latter is the notion generally affirmed by the social democratic ethos, favored by the standard account of bioethics and supported by a liberal cosmopolitan ethos (Engelhardt, 2000, esp. Ch. 3). At stake is a contrast between two competing views of human flourishing, each with its own ethos and settled moral judgments.

Bioethics emerged as a critical response to traditional, largely paternalistic, social structures. Bioethics was made in America.[16] It emerged in order to fill a moral vacuum of public policy leadership created by a number of developments, including (1) the deprofessionalization of medicine,[17] which displaced physicians as the source of authority for health care policy, as well as (2) the secularization of American culture, which displaced religious leaders as guides for public medical morality.[18] These cultural changes in the character of mid-20th century America were integral to a larger social revolution, which challenged traditional authority figures, including physicians. It led to (3) the replacement of the professional standard for the disclosure of information for informed consent by the objective or reasonable-and-prudent-person standard.[19] Bioethics has been a movement with a secular moral and social agenda that, among other things, recognizes individual autonomy or liberty as a principle lexically prior to concerns with family structure, as well as persons actually or hypothetically considered as the source of secular moral authority (Engelhardt, 1986). This bioethics was directed against traditional paternalistic practices, which gave physicians and family members authority over adult patients, including those who had not lost decisional capacity.[20] The result was the establishment of a reigning bioethical ideology defining the character of acceptable moral understandings regarding medical decision-making primarily in individualistically oriented terms.[21] Given the commitment to social and moral reform, it carried within it an anti-traditional animus.

This standard account holds that physicians should treat adult patients as independent sources of authority for initiating or refusing treatment in isolation from control by their families and without the paternalistic overreaching of physicians. Most significantly, personal choice is not only recognized as a source of authority but in addition considered a cardinal source of value: individual, self-directed choice is held to be integral to individual identity and to human flourishing. Even when consented to, paternalistic interventions by families or physicians are in this light interpreted as undermining the respect due to the patient, as well as failing to allow patients to develop their own powers of self-determination and self-realization. The emphasis is on a self-directed and self-regarding understanding of the human good that gives accent to self-development and self-satisfaction.

In summary, this individually oriented account of health care decision-making comports with the commitments of a liberal, social-democratic view of individuals and the family, which can procrusteanly and for heuristic purposes be starkly summarized as follows.

A. The social ontology of the family must be accounted for in terms of the individuals that produce it and their goals: individuals create the authority and structure of families contractually
   1. individuals are regarded as the original source of authority for treatment decisions primarily involving them
   2. individual, autonomous determination, self-fulfillment, and self-realization are appreciated as valuable in themselves and integral to a morally authentic life
   3. families are seen as primarily consensually fashioned social structures, as social creations through the free choices of individuals who decide to collaborate in sexual and reproductive undertakings in the pursuit of a cluster of freely endorsed goals.

> 4. this account is one that involves a methodological individualism, which regards authority and values as in the end to be understood in terms of and reduced to individuals and their interests.
> 5. as a consequence of the commitment both to individual autonomy and individual self-realization, traditional patriarchal family structures are understood as appropriately to be criticized, reshaped, reformed, and marginalized.
> 6. as a consequence as well, this individually oriented, liberal cosmopolitan account of the family does not recognize as normative traditional heterosexual family structures.

B. This account of the family possesses important implications for clinical decision-making and public policy

> 1. the authorization for the treatment of competent adult patients must be derived from the patients themselves, the ultimate source of authority
> 2. the autonomous choice of patients about their own treatment is to be encouraged as integral to human flourishing
> 3. third parties should be allowed to be involved in the treatment of their family members through their receiving medical information concerning the patient and/or acting as surrogate decision-makers, only insofar as this is explicitly authorized by the patient
> 4. the burden of proof is on family members who wish to be involved in a patient's treatment choices to demonstrate that the patient authorizes or has authorized their involvement
> 5. patients who decline active involvement in their treatment choices in favor of their family members assuming authority should be carefully questioned in order to protect them against illicit third-party pressure and manipulation in order to support authentically autonomous individual choice (i.e., this individually oriented ethos involves an individualistically oriented bioethics of suspicion)
> 6. the state, health care institutions, and health care professionals should protect individuals from abuse by, inordinate manipulation from, and intrusion on the part of, family members

In light of this account of the centrality of individuals and of the moral significance of individual choice, health care policy should focus on assuring that patients are empowered to act as the stewards of their own destinies. This context makes plausible both (1) allowing the re-fashioning of family structures in new ways on the basis of consent and through the aid of technological intervention, as well as (2) reducing the meaning of the family to its significance, for and with respect to, its members.

IV. FAMILIST ACCOUNTS: AUTONOMY AND AUTHORITY OF THE FAMILY AS A SOCIAL CATEGORY

Individually-oriented accounts of moral conduct and appropriate bioethical choice contrast with many traditional family-oriented accounts of moral choice, especially those that recognize families as intermediate social institutions between the individuals and the state. In the latter accounts, families are presumptively considered to be the source of authority for decisions to accept or refuse treatment on behalf of their members. In its full presentation, somewhat procrusteanly and starkly put for heuristic purposes, a family-oriented account of health care decision-making compasses the following elements.

A. The social ontology of individuals and the family requires accepting families as independent sources of moral authority: families create and sustain the moral and social persona of their members (i.e., the being of individuals is social, and in particular familial). In such accounts,

1. families are regarded as social structures intermediate between individuals and the state, possessing a *sui generis* authority over their members; families are somewhat like weak mini-states;

2. families are valued, not just because of the benefits they confer to their members, but most importantly because they bring into existence an independent cluster of values integral to human flourishing;

3. families are regarded as sustaining an irreducible, morally significant web of relationships, which constitutes a genre of mutual recognition valued in itself as integral to the moral life.

4. this account is one that involves a methodological familism that recognizes the family as an emergent social reality with emergent social values and structures irreducible to individuals and their interests.

5. as a consequence of the commitment both to family authority and the realization of family flourishing, traditional patriarchal family structures are not brought into question but in general are affirmed and nurtured.

6. as a consequence as well of this family- and traditionally-oriented account, heterosexual family structures are accepted as normative.

B. This account of the family possesses implications for clinical decision-making

1. authorization of the treatment of competent adult patients should as far as possible be acquired through discussions with the family, not just the patient

2. family-oriented patient choices should be encouraged as integral to human flourishing

3. the family should be expected to be involved in the treatment of its members through receiving medical information concerning those members and/or acting as their surrogate decision-makers, unless this is
   a. explicitly rejected by the patient or
   b. clearly conflicts with the patient's best interests

4. the burden is on individuals who do not wish their family to be involved in their treatment decisions to exit from the presumptive family-oriented context of medical decision-making

5. individuals who fail to involve their family in medical decisions should be carefully questioned to protect them from the isolation, anomie, and loss of meaning associated with individualistic models of medical decision-making;

they should be protected against the moral distortions of a dominant individualistic ideology (i.e., there is a family-oriented bioethics of suspicion)
6. the state, health care institutions, and health care professionals should act to protect families from abuse, inordinate manipulation by, and intrusion into their private sphere by individuals. The autonomy of families should be protected.

Such family-oriented approaches to health care decision-making in part draw on traditional cultural understandings of the significance and importance of the family and in part are reactions against the social anomie of contemporary Western societies in which individuals are increasingly left to fend for themselves with fewer robust mediating social structures between them and the state.[22] It is not simply that Western Europeans are ever more not reproducing, but families find themselves broken and dispersed by geography, divorce, various forms of alienation, and a loss of proper structure, as well as appropriate goals. The concern to avoid this state of affairs likely plays a role as well.

Concerns with family integrity have an analogy with those concerns for institutional ethics that recognize institutions as corporate moral persons. Not only can institutions be regarded as legal persons, but as institutional or social persons that can be appreciated as responsible or irresponsible, as possessing or failing to maintain integrity.[23] Methodological familism, the commitment to regard families as integral social realities irreducible to their members and as in possession of interests and goods that transcend individual interests and goods, invites the exploration of a range of issues, from family identity to family autonomy and integrity. In this context, any use of the new reproductive technologies is seen to be morally constrained by the social ontology of the family, as well as the ethics of family integrity and responsibility.

After a period of uncritical acceptance, it is only to be expected that there would be critical reassessment of the standard individualist bioethical account of the new reproductive technologies and their implications for the family.[24] What is significant is that a fundamentally different account of medical decision-making can be recognized that acknowledges family autonomy and authority, the family as a social space and category necessary for the realization of full human flourishing, and traditional views of proper conduct and family structure. The current controversies and battles of the culture wars associated with the family thus involve more than the various challenges to traditional structures, goals, and values that are intensified by the new technologically mediated recastings of the expected patterns of human sexuality and reproduction. The controversies derive as well from disputes regarding the categorial or ontological status of the family and its appropriate moral and legal authority.

V. THE PERSISTENCE OF FAMILY AUTHORITY

Although on the surface, it may appear as if in the West the authority of the family has been largely reduced to that of its constitutive authorizing individuals, this is far from clear. For example, despite the American cultural endorsement of the standard individualistic account of appropriate bioethical decision-making, there is good evidence that bioethical decision-making, at least in many American jurisdictions, may in its day-to-day operation be closer to many of the familist practices of Asia than might at first examination be suspected. For example, although at the surface level bioethics and health care policy in the United States often accords with the commitments of the standard individualistic American account of bioethics, space is frequently afforded for a largely hidden, operatively familist bioethics.[25] The extent of this hidden or sub-surface ethics set over against the manifest ethos and its morality then is by necessity somewhat conjectural, and will require for its assessment further careful empirical studies of actual decisions by actual physicians, patients, and families.[26] Also underexamined is the extent to which such a two-tier or two-dimensional approach to morality and public policy allows both (1) an accommodation to moral pluralism and (2) a lessening of overt moral controversy by seeming acceptance of the dominant bioethical account, while at the same time allowing traditional moral commitments to continue in place.

An indication of the presence of a two-tier approach to morality is found in statutory provisions regarding medical decisions on behalf of previously competent individuals who no longer possess decisional capacity. For instance in Texas, individuals who once had possessed decisional capacity are often placed within a framework of authority structured by familist presuppositions. Texas Statutes provide that, in the absence of an agent appointed through a medical power-of-attorney[27] or a court-appointed guardian, the following persons in the following order are in authority as surrogate decision-makers for all except a small set of medical decisions:[28]

1. the patient's spouse;
2. an adult child of the patient who has the waiver and consent of all other qualified adult children of the patient to act as the sole decision-maker;
3. a majority of the patient's reasonably available adult children;
4. the patient's parents; or
5. the individual clearly identified to act for the patient by the patient before the patient became incapacitated, the patient's nearest living relative, or a member of the clergy.[29]

When these individuals make treatment decisions, the statute requires that "Any medical treatment consented to under Subsection (a) must be based on knowledge of what the patient would desire, if known."[30] A similar order of authority is recognized at law for the withdrawal of life-saving treatment[31] from patients who lack decisional capacity and who are suffering from a terminal condition[32] or who have what the statute terms an "irreversible condition".[33]

Two points are of particular significance. First, such legislation allows an easy recognition of family members as surrogates for medical decision-making, including end-of-life decision-making. Second, the legal standard by which the claims of the surrogate decision-maker are accepted as authoritative involves the least exacting test for truth. In the United States, three standards for truth are employed at law:

(1) the first requires that evidence establish the truth of a claim beyond a reasonable doubt (i.e., the test used in criminal law); (2) the second requires that the evidence establish the claim's truth clearly and convincingly (the criterion used for the long-term commitment of mentally ill persons who are dangerous to themselves and/or others); (3) the third requires that a claim be supported by a preponderance of proof (the threshold of evidence used for the civil recovery of damages). This last test demands no more than that the evidence in favor of any claim accepted be slightly more than fifty percent. In terms of this last test of truth, the plausibility of claims made by the family surrogate decision-maker accurately to reflect the past wishes of the patient need only barely outweigh evidence that this claim is implausible.[34]

The result is that Texas informed consent law with respect to decision-making on behalf of an incompetent patient establishes considerable space for family autonomy and authority in that reports by relevant family members regarding the patient's previous wishes are considered to be truthful and therefore authoritative until proven otherwise. Should family members voice their own treatment preferences for an incompetent patient, such family members are then informed that medical decisions may not simply reflect the surrogate's wishes, but should instead reflect the patient's wishes. In many if not most circumstances, this instruction may be materially equivalent to coaching the family surrogate decision-maker to phrase treatment choices in terms required by the standard bioethical norms. That is, the family surrogate decision-makers may then simply aver that their wishes as a family member are also the wishes of the patient. In such circumstances, the *de facto* outcome of the application of the law as currently written and interpreted is generally to place patients within the authority of their family unless those patients have previously taken steps while competent to exit the grasp of family authority through procedures such as appointing a proxy or agent by means of a medical power-of-attorney.[35] Health care decision-making in many jurisdictions in the United States turns out to possess a more traditionally family-oriented character than might at first blush appear to be the case.[36] The same appears as well to be the case in other jurisdictions (Chan, 2004).

Although this exploration of health care decision-making does not focus on the engagement of the new reproductive technologies, it does involve decision-making in the context of high-technology medicine, including decisions regarding the continuance of critical care and high technologically mediated physiological support, where these may only prolong or postpone death. This state of affairs leads individuals, families, and societies to the question of the appropriate use of such high technology and to the issue of whether such decision-making should be construed in primarily individualistic terms. Given the background, dominant account of bioethics, the availability of high-technology medicine in general and with respect to end-of-life decision-making would suggest that family-oriented authority should be marginalized and individuals placed robustly in authority. The evidence indicates that matters are much more nuanced and that places for family-oriented decision-making have been preserved.

VI. CONFLICTING VISIONS OF THE FAMILY, THE CULTURE WARS, AND THE NEW REPRODUCTIVE TECHNOLOGIES: A CONCLUDING REFLECTION

Matters regarding the standing of families are thus very complex. On the one hand, the traditional family and its authority still play a role in society as well as health care, law, and policy. Yet, from contraception, abortion, the use of donor gametes, *in vitro* fertilization, and gestational surrogate motherhood to human reproductive cloning, the new reproductive technologies in various ways separate sexual activity, reproduction, and gestation, activities whose particular unity in the family has framed human experience for millennia. The recent technologically mediated interventions of reproductive medicine and the genetic technologies have brought into question both traditional family structures, as well as the family as an independent category of social reality. These interventions do so by making family structures an object for willful restructuring by collaborating, consenting individuals. It is not simply that new structures and relationships can be fashioned, but that this increasing range of technologically mediated choices suggests both that such choices are licit, and that the family is merely the creature of individual agreements and peaceable collaboration. The intersection of these two issues and the controversies they engender, namely, controversies regarding (1) the morally appropriate ways to interrelate sexual activity, reproduction, and gestation, as well as with regard to (2) the categorial or ontological significance and moral authority of the family, are now being seriously re-examined.[37] Because these issues bear on such powerfully emotively freighted human undertakings as sexuality, reproduction, and the nurturing of children, they tend to generate bitter controversies.

These disputes would be bad enough as they are. They are, however, as this essay has noted, further complicated by the presence of new reproductive and genetic technologies that allow the involvement of numerous third parties in reproduction, as well as promising the possibility of designing one's own children, thus transforming children into the willful products of their parents. In summary, the presence of these technologies makes three cardinal clusters of issues salient:

A. the character of the appropriate human aims or purposes to be pursued through these new technologies. The new technologies make more salient the divide between those who have a thick, traditional, communal understanding of the meaning and purpose of the family and its place in human flourishing over against those who see the only final constraints to be human free choice (after all, to speak of costs and benefits one must first agree about how to weight and rank particular costs, benefits, risks, etc.);

B. the scope of appropriate or proper human willful choice in restructuring the structure of the family. The new technologies make the old problems regarding appropriate individual choice more pressing; and

C. the categorial integrity and independence of the family. The new reproductive technologies suggest that the family is not an emergent reality with an authority of its own, but the free creation of collaborating individuals.

In all of this, it is important to acknowledge the depth of the disagreements between individually oriented and familist oriented accounts of individuals and the place of husband, wife, and children in families. At stake is not simply the status of families, but foundationally different accounts of the human good and human flourishing.

These controversies invite those who live within a traditional moral and social life-world to bring their various traditional cultural and religious understandings of the family to the exploration of the significance and force of their accounts and the commitments they involve. For nearly half a century, the accent public policy has fallen on individually oriented, reductive accounts of the family. Those whose roots are in traditional Christian,[38] Jewish, Islamic, and Confucian[39] presentations of human flourishing and the human good should now turn to the exploration of their own appreciations of the family and the proper uses of our new reproductive technological powers. Such explorations may produce a dialectic of moral and ontological accounts with wide-ranging and surprising implications. Such examinations and the cultural conflicts they portend have likely only begun (Fan and Tao, 2004).

NOTES

[1] Since the Enlightenment in the West, it is not just that natural theological reflection has been marginalized, but religious commitments have come to be considered unessential to the moral life and human flourishing. Whatever may be the case regarding this complex history, it surely is the case that religions have the capacity to unite their members in deep commitment to particular understandings of the family and the nature of human flourishing. See Engelhardt (2000).

[2] Besides accenting autonomy, critical rationality, and liberation from clerical influence, the Enlightenment laid heavy accent on progress not only in science and technology, but in freedom from superstition and the uncritical acceptance of traditional cultural restraints. For example, Immanuel Kant looked forward to the advent of a universal religion of reason.

Pure religious faith alone can found a universal church; for only [such] rational faith can be believed in and shared by everyone, where an historical faith grounded solely on facts, can extend its influence no further than tidings of it can reach, subject to circumstances of time and place and dependent upon the capacity [of men] to judge the credibility of such tidings. Yet, by reason of a peculiar weakness of human nature, pure faith can never be relied on as much as it deserves, that is, a church cannot be established on it alone (Kant, 1960, p. 94, AK VI, 102–103).

Such expectations of progress gave issue to the progressivist aspirations that from the mid-19th century to the early 20th century led to an expectation of universal liberation.

"But the world is growing better. And in the Future—in the long, long ages to come—IT WILL BE REDEEMED! The same spirit of sympathy and fraternity that broke the black man's manacles and is today melting the white woman's chains will tomorrow emancipate the working man and the ox; and, as the ages bloom and the great wheels of the centuries grind on, the same spirit shall banish Selfishness from the earth, and convert the planet finally into one unbroken and unparalleled spectacle of PEACE, JUSTICE, and SOLIDARITY" (Moore, 1906, pp. 328–29).

Rather than the 20th century becoming an age of peace and justice, it was marked by the slaughter of millions in the name of justice and equality.

[3] The failure of rational discursive reflection (e.g., philosophy) conclusively to provide the foundations for a canonical moral account has been recognized for over two millennia. This difficulty is appreciated in the observation by Protagoras, "that there are two sides to every question, opposed to each other" (Diogenes Laertius, 1931, p. 463). If one grants different initial premises, one will produce different and often contradictory conclusions. This state of affairs was widely recognized in the ancient world, not

only by Sophists, but by Christians. Clement of Alexandria, for instance, observes, "Should one say that Knowledge is founded on demonstration by a process of reasoning let him hear that first principles are incapable of demonstration; for they are known neither by art nor sagacity" (Clement of Alexandria, 1994, Book 2, Chapter IV, vol. 2, p. 350). The point is that there has been a long and robust recognition of the limits of discursive rationality, leading to the conclusion that it is impossible to establish through sound rational argument a particular moral account without begging the question, arguing in a circle, or engaging an infinite regress. The recognition of a five-point basis for the inconclusiveness of many philosophical arguments found a classic articulation in the *pente tropoi* or five modes of the later Skeptics and the late Academy. Diogenes Laertius (3rd century A.D.) attributes these to a Greek Skeptic, Agrippa. "But Agrippa and his school [affirm five] modes, resulting respectively from disagreement, extension *ad infinitum*, relativity, hypothesis and reciprocal inference" (Diogenes Laertius, 1931, vol. 2, p. 501, IX.88). A similar account is provided by Sextus Empiricus (probably third century) of the five modes: "the first based on discrepancy, the second on regress *ad infinitum*, the third on relativity, the fourth on hypothesis, the fifth on circular reasoning" (Sextus Empiricus, 1976, Vol. 1, p. 95, I.164).

[4]   The term culture wars gained general currency by Hunter (1991). See, also, Huntington (1997).

[5]   For an example of an individually oriented, community-directed account, see Rawls (1971), who uses the notion of an original contractarian perspective as an expository device in order to impose a particular moral perspective. The later Rawls nests this normative perspective within a social democratic commitment to a constitutional polity. See Rawls (1993). In each case, the result is a communitarian account that is also individually oriented. In one of his last essays, Rawls critically assesses the independence of the family as a private domain of moral choice. Rawls argues that "the spheres of the political and the public, of the nonpublic and the private, fall out from the content and application of the conception of justice and its principles. If the so-called private sphere is alleged to be a space exempt from justice, then there is no such thing" (Rawls, 1997, p. 791). Among other things, Rawls affirms the reshaping of traditionally structured families. "Mill held that the family in his day was a school for male despotism: it inculcated habits of thought and ways of feeling and conduct incompatible with democracy. If so, the principles of justice enjoining a reasonable constitutional democratic society can plainly be invoked to reform the family" (Rawls, 1997, pp. 790–91).

[6]   "Traditional family" is used (1) to identify a permanent, patriarchally structured, sexual and repro-ductive alliance of a man and a woman, as well as (2) to recognize this structure as both (a) possessing an autonomy and authority and (b) constituting an integral element of human flourishing.

[7]   For an exploration of the ways in which families are stumbling-blocks to fair equality of opportunity, see Fishkin (1983).

[8]   One must note Islam's acceptance of polygamy as an important traditional family pattern.

[9]   Although there are numerous portrayals of the Christian account of the family, one is scripturally rooted and persists as the core of the tradition. See I Cor 11:11–16; Eph 5:23–33; and Col 3:18–20.

[10]   See, for example, St. John Chrysostom's Homily XII on Colossians iv.12–13.

[11]   It is important to underscore that there is no single Confucian or other Asian account of the family. All Confucian accounts require selecting particular writings from a rich history and literature or endorsing particular persisting traditional patterns as normative or canonical. To give a Confucian account requires a normative reconstruction of the tradition, as well as persisting patterns, so as to establish one account as guiding. See, for example, Fan (2002).

[12]   An exploration of the dramatic impact of effective contraception on the family and social structures generally is given by Fukuyama (1999).

[13]   In the United States and Canada, as well as in Western Europe, there has been a fundamental reshaping of cultural and legal presuppositions regarding appropriate human sexuality and reproduction. See, for example, *Lawrence v. Texas*; *Goodridge v. Dept. of Public Health*; *In the matter of Section 53*; and *In the matter of a Reference*.

[14]   Ruiping Fan provides a proposal for a reconstructionist Confucianism that can account for the autonomy and integrity of the family. See, for example, Fan (2002 and 2003).

[15]   Hegel provides a categorial account of social structures, including the family. See, for example, G. W. F. Hegel, *Grundlinien der Philosophie des Rechts*, especially §§ 158–181. For an overview of this approach to social reality, see Hartmann (1976).

[16]   The term bioethics was coined by Van Rensselaer Potter (1970a, 1970b, 1971). Potter sought to engender an ethos aimed at preserving the biosphere (Potter, 1988). "Bioethics" appears to have been either independently coined or fundamentally recast in its meaning by Sargent Shriver and/or André Hellegers (Sargent Shriver, letter to author, January 26, 2001). See also Reich, 1994. An overview of the development of bioethics as the new health care morality is provided by Jonsen (1998).

[17]   During the mid-20th century, in various steps American medicine was transformed from a quasi-guild to a trade. See *USA v. AMA*; and *AMA v. Federal Trade Comm'n*. See also Starr (1982).

[18]   The United States at the beginning of the 20th century were *de jure* a Christian nation. See, for example, *United States v. Macintosh*. In mid-century, the United States were transformed into a *de jure* secular polity. See, for example, *Tessim Zorach v. Clauson* and *School District v. Schempp*.

[19]   The paternalist role of the physician was brought into question, *inter alia*, through the establishment of the so-called objective or reasonable-and-prudent-person standard for disclosure in informed consent. This standard required disclosing that information material for a reasonable and prudent patient to accept or refuse a particular medical intervention (*Canterbury v. Spence*).

[20]   Early bioethical criticisms of paternalism include Fried (1974); Annas (1975); Buchanan (1978); Dworkin (1972).

[21]   These matters are explored at greater length in Engelhardt (2002).

[22]   The standard American account of bioethics is increasingly being subjected to critical reassessment; see Hoshino (1997). Alora and Lumitao (2001) offer an interesting account of a familist bioethics.

[23]   For a non-reductive exploration of institutional authority and integrity, see Iltis (2005).

[24]   One expression of a reaction against a liberal cosmopolitan bioethics is found in George W. Bush's President's Council on Bioethics, 2002, 2004, and 2003. See, also, an example of the reaction against the dominant bioethics in Smith (2000, chapter 3).

[25]   For a further exploration of some of these issues, see Cherry & Engelhardt (2004).

[26]   Studies of this underground familist bioethics will be difficult, given the strategically hypocritical or clandestine character of such decision-making. In the face of a dominant liberal individual ethos, familism will tend to be clandestine. After all, if the system functions by providing a surface endorsement of the standard American account, while at the same time in a hidden fashion operating in a familist mode, the very logic of this state of affairs requires never frankly acknowledging the presence of the hidden, albeit operative, familist bioethics. For a ground-breaking empirical study of these matters, see Boisaubin (2004).

[27]   Texas Statutes, 166.

[28]   This order of the authority of surrogate decision-makers does not apply to decisions to withdraw or withhold life-saving treatment on a person with a terminal or irreversible condition, nor does the law allow surrogate consent to electroconvulsive therapy.

[29]   Texas Statutes, 313.004.(A).

[30]   Texas Statutes, 313.004.(c).

[31]   Texas Statutes, 166.002.(10): "'Life-sustaining treatment' means treatment that, based on reasonable medical judgment, sustains the life of a patient and without which the patient will die. The term includes both life-sustaining medications and artificial life support, such as mechanical breathing machines, kidney dialysis treatment, and artificial nutrition and hydration. The term does not include the administration of pain management medication or the performance of a medical procedure considered to be necessary to provide comfort care, or any other medical care provided to alleviate a patient's pain."

[32]   Texas Statutes, 166.002.(13): "'Terminal condition' means an incurable condition caused by injury, disease, or illness that according to reasonable medical judgment will produce death within six months, even with available life-sustaining treatment provided in accordance with the prevailing standard of medical care. A patient who has been admitted to a program under which the person receives hospice services provided by a home and community support services agency licensed under Chapter 142 is presumed to have a terminal condition for purposes of this chapter."

[33]   Texas Statutes, 166.002.(9): "'Irreversible condition' means a condition, injury, or illness: (A) that may be treated but is never cured or eliminated; (B) that leaves a person unable to care for or make decisions for the person's own self; and (C) that, without life-sustaining treatment provided in accordance with the prevailing standard of medical care, is fatal." The last condition includes irreversible afflictions that leave the patient unable to care for himself or make decisions on behalf of himself and who

in addition would die without life-saving treatment. This latter definition embraces patients in such conditions as permanently vegetative states, as well as the advanced stages of Alzheimer's. Among the treatments that may be withheld or withdrawn are artificial nutrition and hydration. Texas Statutes, 166.002.(10). The previous statute similarly provides that the surrogate (in this case the patient's spouse, the patient's reasonably available adult children, the patient's parents, or the patient's nearest living relative [Texas Statutes § 166.0395]) is asked to make decisions "based on knowledge of what the patient would desire, if known." Texas Statutes, 166.039.(c).

[34] The test of truth as a preponderance of evidence for claims by families regarding the wishes of their members is not recognized in all jurisdictions, not even in all American jurisdictions. A number of U.S. jurisdictions employ instead the clear and convincing evidence test, as acknowledged by the United States Supreme Court holding in the Cruzan case with respect to Missouri (*In re Cruzan*). See also *Wendland v. Wendland*. Disputes over these issues in New York produced a particularly productive debate. See Baker and Strosberg (1995). Jurisdictions such as Germany have been so concerned to protect patients against the family's overreaching that even family members appointed under a durable power-of-attorney must in addition secure court approval for any therapeutic interventions associated with a significant risk of death or permanent damage to health. Bundesgesetzbuch, § 1904. For some critical reflections on this law, see 'Sterben und Tod' (Arbeitsgruppe, 1998).

[35] One might conclude that a complex adaptation has been secured so that, as long as lip service is given to the standard bioethics at the surface level, a family-oriented approach can nevertheless cryptically govern, leading to a distinction between the manifest and functioning ethos with its different moral commitments.

[36] Texas law acknowledges the authority of physicians when deciding on behalf of patients. It provides that, in the absence of a guardian, an agent appointed under a medical power-of-attorney, or an authorized surrogate decision-maker, two physicians may make decisions to withhold or withdraw treatment from terminal patients who lack decisional capacity, as well as from patients in an irreversible condition. "If the patient does not have a legal guardian and a person listed in Subsection (b) is not available, a treatment decision made under Subsection (b) must be concurred in [witnessed] by another physician who is not involved in the treatment of the patient or who is a representative of an ethics or medical committee of the health care facility in which the person is a patient." Texas Statutes, 166.039.(e).

[37] Reassessments of family authority are now occurring. See a recent international conference, "The Role of the Family in Medical Decision Making," Evangelische Akademie Arnoldshain, June 8–9, 2004, sponsored by Forum für Ethik in der Medizin e.V. (Frankfurt/M.), Zentrum für Ethik in der Medizin am Markus-Krankenhaus (Frankfurt/M.), and Professur für Strafrecht und Strafprozessrecht, Universität Gießen.

[38] A study of the traditional Christian conception of family structure, sexuality, and reproduction is provided in Engelhardt (2000), especially chapter 5.

[39] For an exploration of the presuppositions involved in a traditional Confucian appreciation of the family, see Cong (2004); Fan and Li (2004); Tse and Tao (2004).

REFERENCES

Alora, A.T. & Lumitao, J.M. (Eds.) (2001). *Beyond a Western Bioethics*, Georgetown University Press, Washington, DC.

American Medical Assoc. v. Federal Trade Comm'n, 638 F.2d 443 (2d Cir. 1980).

Annas, G.J. (1975). *The Rights of Hospital Patients*, Avon, New York.

Arbeitsgruppe 'Sterben und Tod' der Akademie für Ethik in der Medizin (1998). 'Patientenverfügung, Betreuungsverfügung, Vorsorgevollmacht,' Akademie für Ethik in der Medizin, Göttingen, November.

Baker, R.B., Bynum, J., & Strosberg, M. (Eds.) (1995). *Legislating Medical Ethics*, Kluwer Academic Publishers, Dordrecht.

Boisaubin, E.V., Jr. (2004). 'Observations of Physician, Patient, and Family Perceptions of Informed Consent in Houston, Texas,' *Journal of Medicine and Philosophy*, 29 (2), 225–236.

Buchanan, A. (1978). 'Medical Paternalism,' *Philosophy & Public Affairs*, 7 (Summer), 370–390.

Canterbury v. Spence, 464 f.2d 772 (DC Cir 1972).

Chan, H.M. (2004). 'Informed Consent Hong Kong Style: An Instance of Moderate Familism,' *Journal of Medicine and Philosophy*, 29 (2), 195–206.

Cherry, M.J. & Engelhardt, H.T., Jr. (2004). 'Informed Consent in Texas: Theory and Practice,' *Journal of Medicine and Philosophy*, 29 (2), 237–252.

Clement of Alexandria (1994). 'The Stromata,' in A. Roberts, J. Donaldson, P. Schaff, & H. Wace (Eds), *Ante-Nicene Fathers*, Vol. 2, Hendrickson Publishers, Peabody, MA.

Cong, Y. (2004). 'Doctor-Family-Patient Relationship: The Chinese Paradigm of Informed Consent,' *Journal of Medicine and Philosophy*, 29 (2), 149–178.

Diogenes, L. (1931). *Lives of Eminent Philosophers*, R.D. Hicks, (trans.), Vol. 1-2, Harvard University Press, Cambridge, MA.

Dworkin, G. (1972). 'Paternalism,' *Monist*, 56, 64–84.

Engelhardt, H.T., Jr. (1986). *The Foundations of Bioethics*, (Revised Edition. 1996), Oxford University Press, New York.

Engelhardt, H.T., Jr. (2000). *The Foundations of Christian Bioethics*, Swets & Zeitlinger, Lisse, Netherlands.

Engelhardt, H.T., Jr. (2002). 'The Ordination of Bioethicists as Secular Moral Experts,' *Social Philosophy & Policy*, 19 (Summer), 59–82.

Fan, R. (2002). 'Reconstructionist Confucianism and Bioethics: A Note on Moral Difference,' in H.T. Engelhardt, Jr., and L.M. Rasmussen (Eds.), *Bioethics and Moral Content: National Traditions of Health Care Morality* (pp. 281–287), Kluwer Academic Publishers, Dordrecht.

Fan, R. (2002). 'Reconstructionist Confucianism and Health Care: An Asian Moral Account of Health Care Resource Allocation,' *Journal of Medicine and Philosophy*, 27 (6), 675–684.

Fan, R. (2003). 'Rights or Virtues? Toward a Reconstructionist Confucian Bioethics,' in R. Qiu (Ed.), *Bioethics: Asian Perspectives* (pp. 57–68), Kluwer Academic Publishers, Dordrecht.

Fan, R. & Benfu, L. (2004). 'Truth Telling in Medicine: The Confucian View,' *Journal of Medicine and Philosophy*, 29 (2), 179–193.

Fan, R. & Tao, J. (2004). 'Consent to Medical Treatment: The Complex Interplay of Patients, Families, and Physicians,' *Journal of Medicine and Philosophy*, 29 (2), 139–148.

Fishkin, J.S. (1983). *Justice, Equal Opportunity, and the Family*, Yale University Press, New Haven, CN.

Fried, C. (1974). *Medical Experimentation*, American Elsevier, New York.

Fukuyama, F. (1999). *The Great Disruption: Human Nature and the Reconstitution of Social Order*, Free Press, New York.

Goodridge v. Department of Public Health, 309 (2003) (Mass. Sup Ct).

Hartmann, K. (1976). 'Die ontologische Option,' in K. Hartmann (Ed.), *Die ontologische Option* (pp. 1–30), Walter de Gruyter, Berlin.

Hoshino, K. (Ed.) (1997). *Japanese and Western Bioethics*, Kluwer Academic Publishers, Dordrecht.

Hunter, J.D. (1991). *Culture Wars: The Struggle to Define America*, Basic Books, New York.

Huntington, S.P. (1997). *The Clash of Civilizations and the Remaking of World Order*, Touchstone, New York.

Iltis, A.S. (2005). *Institutional Integrity: A Critical Re-examination*, Kluwer Academic Publishers, Dordrecht.

In re Cruzan, 58 LW, 4916 (June 25, 1990).

In the matter of a Reference by the Governor in Council concerning the Proposal for an Act respecting certain aspects of legal capacity for marriage for civil purposes, as set out in Order in Council P.C. 2003-1055, dated the 16 day of July, 2003.

In the matter of Section 53 of the Supreme Court Act, R.S.C. 1985, Chap. S–26.

Jonsen, A.R. (1998). *The Birth of Bioethics*, Oxford University Press, New York.

Kant, I. (1960). *Religion Within the Limits of Reason Alone*, T. Greene and H. Hudson (trans.), Harper, New York.

Lawrence v. Texas, 2002, U.S. LEXIS, 8680 (U.S. Dec. 2, 2002).

Moore, J.H, (1906). *The Universal Kinship*, George Bell, London.

Potter, V.R. (1970a). 'Bioethics, the Science of Survival,' *Perspectives in Biology and Medicine*, 14, 127–53.

Potter, V.R. (1970b). 'Biocybernetics and Survival,' *Zygon*, 5, 229–46.

Potter, V.R. (1971). *Bioethics, Bridge to the Future*, Prentice-Hall, Englewood Cliffs, NJ.

Potter, V.R. (1988). *Global Bioethics*, Michigan State University Press, East Lansing.

President's Council on Bioethics (2002). *Human Cloning and Human Dignity*, Public Affairs, New York.

President's Council on Bioethics (2003). *Beyond Therapy*, Dana Press, Washington, DC.

President's Council on Bioethics (2004). *Monitoring Stem Cell Research*, President's Council on Bioethics, Washington, DC.

Rawls, J. (1971). *The Theory of Justice*, Harvard University Press, Cambridge, MA.

Rawls, J. (1993). *Political Liberalism*, Columbia University Press, New York.

Rawls, J. (1997). 'The Idea of Public Reason Revisited,' *The University of Chicago Law Review*, 64 (Summer), 765-807.

Reich, W. (1994). 'The Word 'Bioethics': Its Birth and the Legacies of Those who Shaped its Meaning,' *Kennedy Institute of Ethics Journal*, 4, 319–336.

School District of Abington Township v. Edward L. Schempp et al., William J. Murray et al. v. John N. Curlett et al., 374 US 203, 10 L ed 2d 844, 83 S Ct 1560 (1963).

Sextus Empiricus (1976). 'Outlines of Pyrrhonism,' in R.G. Bury (trans.), *Sextus Empiricus*, Harvard University Press, Cambridge, MA.

Smith, W.J. (2000). *Culture of Death: The Assault on Medical Ethics in America*, Encounter Books, San Francisco.

Starr, P. (1982). *The Social Transformation of American Medicine*, Basic Books, New York.

Tessim Zorach v. Andrew G. Clauson et al., 343 US 306, 96 L ed 954, 72 S Ct 679 (1951).

Tse, C. & Tao, J. (2004). 'Strategic Ambiguities in the Process of Consent,' *Journal of Medicine and Philosophy*, 29 (2), 207–223.

United States of America, Appellants, v. The American Medical Association, A Corporation; The Medical Society of the District of Columbia, A Corporation, et al., 317 U.S. 519 (1943).

United States v. Macintosh, 283 US 605 (1931).

Wendland, Robert, Rose Wendland v. Florence Wendland et al., 26 Cal. 4th 519; 28 P.3d 151; 110 Cal. Rptr. 2d 412.

CHAPTER 4

FAMILY LIFE, BIOETHICS AND CONFUCIANISM

STEPHEN A. ERICKSON

Pomona College, Claremont, California, United States

I. LIBERAL SECULAR INDIVIDUALISM AND THE CONFUCIAN RESPONSE TO THE WEST'S ANOMIE: AN INTRODUCTION

In the West a view of life has been gaining adherents with accelerating momentum. I shall label this view *liberal secular individualism*. Not all holders of this view construe themselves as libertarian. Neither do all understand themselves as secular. But the logic of the individualism that drives the West is both libertarian and secular, if it faces up squarely to its own implications. In our twenty-first century it is almost certainly in the domain of biotechnology that these implications will become most starkly visible.

It is misleading, however, to confine this mode of thinking to the West, as opposed to something called the East. Systems of communication, information technologies and market-oriented economics have begun to transform this individualistic mode of understanding into a global reality.

In what follows I will be exaggerating the tendencies inherent in Western individualism through focusing on one distinct biotechnological possibility: the eventual, possibly even mid-twenty-first century, *manufacture* of human beings. But the exaggeration I offer is no more unlikely than wireless communication technologies were to educated people sixty years ago. The manufacture of human beings is a dangerous plausibility that deserves serious attention.

II. MEDICINE AND ITS THREAT TO HUMAN NATURE: TAKING METAPHYSICS AND FAMILY SERIOUSLY

The major danger issuing from a rapidly globalizing biotechnology lies in the relation of medicine to human nature. Human beings are very much *in*, but they are not altogether *of* the world. Until very recently medicine has only spoken to that dimension of human life that is *in* the world, what has traditionally in the West been called the body (soma). Especially in the twentieth century medicine in the

47

S.C. Lee (ed.), The Family, Medical Decision-Making, and Biotechnology, 47–58.
© 2007 *Springer.*

form of psychiatry has also concerned itself with the psyche (*psyche*), but only in terms of human emotional life in the world. In short, medicine has been a worldly undertaking.

But what if human life, however much lived *in* this world, is not altogether *of* this world? What if it is in the nature of human beings to have a relatedness to something that is *beyond* this world? One meaning of the Greek word *meta* is "beyond." From this root we derive the notion of the *metaphysical*. What if human beings are metaphysical in their nature, belonging and also deriving their existence from elsewhere – however much they are also physical beings here, in this world?

Through biotechnology medicine is rapidly colonizing all aspects of human nature. The ultimate program of biotechnology, more latent than explicitly thought through, is to understand *everything* about human nature to the point of reaching the capacity to manufacture particular and specifically designed individual human *natures*. If this result is attained, every aspect of human existence will in principle be producible from within this world and for purposes found solely within the framework of this world. The metaphysical dimension of human nature will have been erased, replaced and/or superceded.

I turn now to some aspects of Confucian thought as they relate to what I am labeling (globalizing) *liberal secular individualism*. Before I do, however, it is important to issue a few cautionary remarks with respect to the sketch I have so far drawn. (1) Not to be altogether *of* the world – in short, to be metaphysical – may be a very complex metaphor. Equally, however, it may indicate a relatedness of the human to something *beyond* the human that humans can neither produce nor control, something quite literally metaphysical rather than physical. (2) In the traditional Western understanding 'psyche' is a boundary word. It straddles the psychological and the spiritual. Perhaps at one extreme all illnesses issuing from the psyche could be experienced as spiritual. From this perspective 'psyche' is best translated as *soul*, and those who minister to it are best construed as spiritual clinicians. The other extreme, however, is the one toward which first psychiatry, then various forms of psychoanalysis, and now biotechnology is leading us. On this account 'psyche' *could* be translated, as I did earlier, simply as itself, as *psyche*. But this is then quickly construed as a shopworn notion that belongs to an outmoded "world view." In its place are to be found the materials discovered through biogenetic investigation. As opposed to spiritual clinicians, those who "minister" to these biogenetic elements are most reasonably bio-pharmacologists and medically trained, neuro-physiologically conversant doctors.

Confucian thought has at least one, very distinct advantage over the spreading individualism issuing from the "West." This advantage exists in separation from the threat to human life that is posed by the biotechnological trajectory I have begun to sketch. The advantage Confucianism offers is a rich and thick understanding of the "family." This Confucian understanding is not easily grasped from an individualistic perspective, however, and must be approached most carefully, if it is not to be distorted.

Two notions are very fundamental to the Confucian vision of human life: *empathy* and *moral agency*. Though they overlap and are inseparably fused, they are nonetheless distinguishable. However controversially, an order of precedence can be established between the two. Agency arises from and is in the intimate service of empathy. Agency might well be termed the incarnation of empathy.

It is precisely in and through families that agencies (persons) first emerge and are nurtured toward "adulthood." In Confucian thought families are the necessary (though arguably not the sufficient) condition for the flourishing of persons. The notion of adulthood, however, often escapes Western, individualistically oriented understandings.

Liberal individualism construes adults not only as fully formed, but as thereby separable and distinct persons. These persons may subsequently enter or have already entered into a variety of human relationships. But these relationships are not constitutive of the core of complete adulthood. Such relations may enormously enrich, but are not internal to the complete adult. They are understood more as clothing than as flesh and bone. The extreme form of this view is anticipated in the seventeenth century in Leibniz's claim that "monads" have no windows.

The Confucian view is quite different. Families are constitutive of persons, equally of those persons who are adults as of those who are children or infants. The family is internal to all parties that enter into its composition. To be "extricated" from one's family is not an emancipation toward personal completion. It is a partial amputation of one's very person.

In this, however, there is a mystery that some may construe as the unspoken threshold of Confucian metaphysics. As the very core of the family, empathy is virtually a unity. It produces duality, trinity, and further plurality. But it remains a *one* that engenders that many that is the family. And it engenders further "many-s" as well. Any attempt to define the bond that is empathy all too easily, perhaps unavoidably falls prey to what may be a necessity of thought: the pre-establishment of a duality for thinking, with empathy subsequently "emerging" as a relation between the duality misleadingly constructed.

I will now contrast Confucian thought and Western liberal secular individualism in a stark, metaphysical manner. I will then turn to less abstract, more historical and contemporary concerns. On this more solid ground what I now convey will not only take on further flesh, but assume a surprising and disturbing degree of urgency.

Western liberal individualism has inherited a tradition that not only connects human existence with life *in* the world, but strongly suggests that human life is not altogether *of* the world. Western liberal individualism has built in and/or discovered the *possibility* of the Transcendent within the assumed actuality of *individual* human being. But over the last few centuries this "Western" openness to Transcendence is being shut down and shut off. There are at least two sources of this closure, one of greater age and the other of far more recent birth. The former is the Enlightenment that began in eighteenth century France, and the latter is the growing research agenda and engineering ambitions of globally oriented biotechnology.

In contrast, Confucian thought has inherited a tradition that first encounters human life within a cradle of familial empathy, intimating but not guaranteeing a larger nexus of bonds. But Confucian thought neither specifically seeks nor seems to require the metaphysical, if by this is meant a beyond, not altogether of this world.

The "West" has been described as axial, bifurcating reality into the immanent and the transcendent, if only and finally to deny Transcendence itself in the name of utopian and relentlessly hubristic dreams of human progress. Life has been construed as a journey from bondage to liberation, confusion to insight, appearance to reality, and darkness to light. In contrast, Confucian thought does not require this axial bifurcation and thereby exhibits greater stability and a temperament oriented toward gradualism. Yet, not having been subjected to axial bifurcation, Confucianism could neither accurately nor fairly be described as a doctrine of immanence. It does not affirm, but neither does it deny nor reject Transcendence. Strikingly, the dynamic of Western rationalistic individualism has led to a void. At its heart, however, Confucian thought is bathed in a sea of real and potential love.

But of course these are condensed and overdrawn contrasts. As of the early twenty-first century, the picture I have painted of the "West" is unduly nihilistic. As I shall soon show, secular and centrifugal forces are inherent in the very rationalism that the West has celebrated. Yet in many ways and in many places in the "West" families and communities flourish. Bankrupt and isolated individualism need not be Western "fate." And the portrait I have presented of Confucian thought is not only highly complementary, but it is also excessively idealistic. Dysfunctional families exist within the Confucian world as well as in the fraying "West." And industrial and technological progress has not only been great in the Confucian world. It has brought with it some measure of hubris, disorientation and uncertainty as well.

III. GLOBALIZATION, AMERICANIZATION, AND THE CARDINAL PLACE OF CONFUCIAN THOUGHT

Almost paradoxically, it is one of the centripetal forces at work in the early twenty-first century that should most concern us. The label given to it has been *globalization*. Globalization has many manifestations. One is the extension of communication networks. Others involve the spread of free markets for the allocation of capital and the exchange of goods and services. Underlying globalization is the notion of productivity enhancement. Its focus is economic, and its goal is the achievement of higher living standards and greater material wellbeing throughout the world.

A number of aspects of these seemingly positive and even salutary developments deserve our closer attention. 'Globalization' may be only a more diplomatic synonym for 'Americanization' and by 'Americanization' is meant that liberal secular individualism I have been outlining. It has some disturbing features. A phrase often used in its context is 'creative destruction.' At the point a new, more efficient instrument or means of transaction is forged, or a better product is created, its predecessor is abruptly abandoned. In the name of rapid progress disruption, discontinuity and the destruction of traditional items and practices rule the day. Such activities are justified as being in the service of more effective and enhancing outcomes. Gradual and consultative deliberations and those subsequent alterations of direction that are based on them – as are embraced within Confucianism – give way to rapid reallocations of resources and alterations, if not revolutions in practice.

A locus of that productivity enhancement that has so driven 'globalization' has been *the individual*. It is through the enterprising capacities of economically daring and creative individuals that the engine of productivity has been fueled. The emigration of individuals at all levels in pursuit of higher standards of living, combined with Capital's search for cheaper labor markets, has made our world both more polyglot and cosmopolitan.

At the same time this has disrupted longstanding traditions, for in the movement of people, enhanced technologies, their products, and the means of their delivery, a global community has begun to form. But it is not actually a community. In numerous geographical regions of the world it involves vast numbers of people, having relocated and living and working side by side, who share little, if anything in the way of culture or values, historical past or even language. Beneath the economically laudatory surface there is cultural and spiritual disorientation and confusion.

The axial schema I introduced earlier is illuminating with respect to these matters. In its terms, though we are *in*, we are not altogether *of* the world. But this statement is obviously incomplete and invites further commentary. Our life in the world grounds us. Whatever ventures or journeys we undertake or are taken up by, arise from this base. Its most fundamental cementing component is the family. To the degree globalization spreads, and its ideology of liberal secular individualism expands its influence, family structures and the very dynamic of the family itself is endangered. Since that dimension of our life that is not altogether *of* the world nonetheless proceeds *from* the world and requires it as its grounding, even this "transcendental" dimension will be severely endangered, distorted and potentially extinguished by familial disruption. Through this growing disruption it will have lost its necessary grounding and some of its essential nutritional elements. This has already begun to happen in the West and its consequences, though largely occluded by the promise of further economic progress, are nonetheless alarming. How the promise of biotechnology contributes to this growing predicament will become evident very shortly.

These outcomes have destructive consequences even within liberal individualism itself, for markets need strong and robust individuals to monitor and profit from them. If such individuals are not actively engaged with markets and their transactions, it is in the inherent dynamic of economic efficiencies that what were meant to be robust individuals will be used, uprooted, disoriented and eventually even damaged and destroyed.

What is it that safeguards individuals from such devastating outcomes, if not families? And what protects families more than those communal traditions that have harbored and bonded them, giving them everyday nourishment in terms of mutually understood and embraced rituals and bonds.

Already in the "West" families have been significantly disrupted by abrupt and rapid economic change. But there is more. The axial model, we know, turns us toward something not altogether of the world, something beyond it. However recessive, it exists within Confucianism as well and is ruled over by the supreme Shang Ti. If families have grounded and nurtured our life in this world, once liberal and secular individualism have rejected this grounding, where does it find its own worldly foundation? And how does it seek, and what has it found to fill the void created – but also left standing – by its own very secularism? If in what may be irreducibly axial terms we remain not altogether *of* the world, what is now to be found in this domain?

Rather than rejecting the existence of such a domain, liberal secular individualism has filled it with promises of unending, market-driven economic progress. In the heart of biogenetics and its biotechnologically driven engineering opportunities, it has even seen itself as hitting the proverbial jackpot: the discovery and development of a highly efficient means of enhancing in radically fundamental ways the abilities of those very individuals who are seeking economic and technological progress.

But there is a worrisome circularity at the heart of this thinking. Economic and technological progress, however transitionally destructive to longstanding customs and traditions, is meant for the purpose of individual enhancement. And individual enhancement is meant to underwrite and serve economic and technological progress. But one must also ask regarding the end, the goal or goals toward which this virtually self-devouring, circular activity is meant to strive. What is it? If there is a meaning to life, what is it?

The very axial bifurcation of human life into this world and a beyond, even if devolved into decaying metaphor, requires that "meaning of life" questions be addressed. Our axial life not only provides a space for such questions. It requires that answers for them be found, content encountered, whether derived from tradition or revelation.

The liberal secular individualism of the "West" has provided its own responses, often unwittingly and without concern for their comprehension. This individualism has led either to an unstated, though in places and at times robustly freedom-loving nihilism – or leads to that strange circularity of self-sustaining and seemingly self-enclosed technological efficiency and enhancement I have just described. These two alternatives are not exclusive, and if the latter receives more open acceptance in the twenty-first century, a strong likelihood, biogenetics will soon become its ultimate instrument.

IV. CRITICAL RATIONALITY, SENTIMENT, AND THE CONFUCIAN NOTION OF THE PERSON

In the meantime the grounding of all human activity, the family, is being uprooted, progressively disregarded and is falling into a state of dysfunction and disrepair. As this mode of "Western" rationalist activity – and the collateral damage it inflicts on its human environment – further globalizes and finds sustenance in cultures external to its own historical dynamic, a twofold result can be expected. Those cultures such as the Confucian that venerate the past and gradual change are in danger of being dislodged from their sustaining traditions and being catapulted into a disorienting future. At the same time this unfortunate occurrence would only further confirm a future that is already experienced in empty, if ambitious ways in the midst of the very individualism that is flooding in from globalizing centers of capital and technology.

Understood through Confucian thought, families differ markedly from their counterparts as experienced through the lenses of the individualistic model. In individualistic terms Confucian families are cumbersome and inefficient. Consider the order of presentation, even, with respect to names. In the former the family name takes precedence. In most Western cases it is the first or given name that has priority, often quickly followed with a nickname: "Just call me Bob." This is of more than sociological interest. In Confucian terms your presentation of yourself is already embedded in a nexus of connections within which you are one of many connecting elements, as much a connect*or* as a connect*ed*. You come forward as part of an interwoven fabric. Even in the fundamental notion of person in Confucianism, *Jen*, what is conveyed is relationship, an intimate relating of compassionate humanity with modes of propriety in behavior and gesture that engender respect.

The dynamic of rational efficiency so prevalent in individualism after the Western Enlightenment finds in this configuring of human life a most ineffective means of promoting task-oriented productivity and, more generally, successful enterprise. What has been required and recruited are separable units, unencumbered by what are viewed as extraneous connections that are directly accountable for specific personal performance. If this is construed by many as impoverishing, it is also reinforced in the West by cults of individual responsibility and genius, and the strongly Christian and Protestant notion that each human soul ultimately stand alone in the presence of God.

What has driven this understanding of human life to the point it has currently reached? And how are biogenetically opened engineering opportunities likely to alter our human future? These two questions are ultimately related.

The West has been unduly influenced by "Enlightenment" doctrines that arose in the eighteenth century. A pivotal enlightenment doctrine is that we must dare to know all that we can. Contrasted with such knowledge is something called superstition: beliefs lacking justification and arising out of powerful and baffling experiences transmitted from a more ancient and primitive past.

The term most frequently contrasted with superstition is *Reason*. Unlike the Confucian notion of *hsin* that fuses head and heart in concentrically-spreading, personalized ways, reason is both depersonalizing and abstract. Enlightenment thinking understands the maturation of the human race as the attainment of greater rationality. The advocates of the Enlightenment value reason as the highest and universal in us, that to which all other human dimensions are to be subordinated.

Here major problems arise. Even sentiment, a Western analogue to the Confucian *hsin* is subsumed under the category of superstition. It is construed as irrational, thus to be overcome, or it is valued to the degree that it conforms to the laws of reason. On this account sentiment becomes an element in mental hygiene, valued primarily for its ability to enhance productivity and mental clarity.

Assumed is that reason is both our nature and our destiny. Sentiment is understood either to be sub-rational, pre-rational or only primitively rational. Even if it is the very glue of concrete human life, a view shared by both Confucianism and Hegel, it is to be managed severely. Enlightenment thinking seeks to overcome sentiment altogether or to develop it to be in full accord with reason's dictates. If in particular cases neither of these options is found to be realistic, sentiment is nonetheless permitted and even in ways embraced in human life, but only as a source of enjoyment or release, not in a decision-determining role or in any manner that might substantially influence policymaking. On this account any weighting given to sentiment would suggest intellectual weakness and fall outside of the boundaries of rigorous and serious bioethical debate, our coming concern.

For enlightenment thinking what is valuable in humans is almost exclusively identified with human reason. Reason becomes the successor notion to the soul. It is in the arena of reason that debates must take place and decisions get taken.

But how is reason itself to be understood? How are we to construe this pervasive driver of recent Western thinking? The most central component in all bioethical reflection is the concept of the human. If the human is more and more circumscribed and even saturated by reason, a closer look at reason itself is required and, especially, at some of its more recent historical manifestations. Reason is far more historical than is usually appreciated.

In the "enlightened" West, Francis Bacon's claim that *knowledge is power* is pivotal. If knowledge is through reason, reason then becomes almost the exclusive instrument – and *only* an instrument – for achieving power. Reason virtually becomes power alone, and humans are increasingly valued (and feared) in terms of the degree of that power they possess and can translate into action. In becoming more and more technical and technological, reason becomes increasingly a force, not the only force, but the one most equipped to dispose over other forces. It is not accidental that a nearly Western word is "productivity."

Once humans are identified with technological reason, either they are understood primarily as powers or they are comprehended as instruments for its achievement. It is in these terms they are valued. Sentiment (*hsin*) also comes to be valued instrumentally for the energy that can be extracted from it or as an instrument enabling its possessors to control further aspects of their environment. These two formulations point to the same, largely unarticulated conclusions: sentiment is taken to be an instrument; human life itself is construed as a means of further empowerment, and reality itself is experienced and understood as energy or power.

We have recently reached a very strange state of things. This becomes more evident when we look at biotechnology, the contemporary embodiment of Western enlightenment rationalism. Considerations of biotechnological interventions with respect to human reproduction – whether in regard to predispositions toward various diseases, the enhancement of diverse faculties, or even the nearly complete "manufacture" of human or neo-humanly analogical lives – growingly come to be understood as empowerment issues. A metaphysics of efficient and effective productivity, based in loci of power traditionally appreciated as infant human life, announces itself. A dark shadow in the background is the very partial, but growing conversion of the notion of person into that of asset or resource.

A most central dimension of what is at stake in our biogenetically influenced future is altogether unique. We can foresee the attempted manufacture of "power centers." We now refer to these power centers as persons or selves, and it is nearly impossible to believe that this discourse will not continue indefinitely. Whether it will continue altogether pervasively will become a problematic issue.

Notions and strategies of empowerment, instrumentation and productive efficiency are emerging and beginning to compete for reflective attention and conceptual space with our traditional person- or self-oriented understandings and considerations. Eventually such power-oriented notions will begin to offer themselves as alternatives and even replacements for person – and family-centered conceptions.

V. THE TIPPING POINT: THE INFANT AS A PRODUCT

Increasing attention is paid today to the notion of a *tipping point*. An obvious example clarifies this concept. At 35 degrees Fahrenheit water does not freeze. Neither does it freeze at 34, or 33 degrees Fahrenheit. But at 32 degrees it does. A tipping point is reached, and a very dramatic change occurs.

It would be somewhat misleading to think that a tipping point will be reached with respect to our conception of *an infant* or human life in general. Yet we have always realized that new additions to the family are also *assets* and *resources*. And later these "resources" become in varying degrees the building blocks for neighborhoods, communities, and even societies. We hope that human lives continue to be treated first and foremost as the lives of *persons*. The familially-grounded notion of *hsin* – in the West empathy or sentiment – has reinforced and defended this hope, but the relentless logic of Western rationalism now threatens it.

What is needed to keep the person and the personal, idealized as *Chun tzu* in Confucianism and as *authentic existence* in the West, in the center of our considerations? To reinforce our sense of *personhood* we need a strengthened notion of *self* that transcends the increasing possibilities of instrumentalist reconstruction and redefinition we face. Genetic modification or enhancement must be understood with respect to a *person*. A person is a *who*, an individual who expresses himself or herself *through* various traits and characteristics, but who is somehow and possibly ineffably also more than the sum of these traits and characteristics. It may well be through "concepts" that we articulate the notion of *self* or *person*. But it is surely through *empathy* or *hsin* that we directly experience and adjust our person-directed concepts and their corresponding intuitions.

The West is not without its own resources. Hegel suggests three components that enter into our human constitution. The first, simply, is *ourselves*, but the other two take us beyond this supposedly simple *self*. The second is *our own conception of ourselves*. It is never in sufficient conformity with the self it conceives. But it is equally a dimension of that self, not made less so through being somehow "inaccurate." The third component is *those introjected (internalized) understandings of how significant others, particularly our family, experiences us*. This third component has a large and often uncomprehended influence on how we in fact experience ourselves, and determine who and what we are. Though it diverges from the first two components, it is equally a dimension of our core humanity. In Hegel as well as in Confucianism the family is its prime mediator.

Under Western influence, productivity and efficiency get construed as the most highly valued features of humankind. Thus people come to be treated increasingly as assets or resources and thereby come to experience themselves more in this manner as well. In what might be called the "post-modern" American marriage, for example, two career economic partnerships are increasingly prevalent. A tipping point is reached when, increasingly, sexually initiated and partially sustained economic partnerships turn toward marriage only as a more economically and societally efficient and productive means of advancement. Though these developments occur through decisions made by specific couples, the social order in which these couples grow up and come to discover and define their options itself begins to tip.

In the partnerships now under discussion, two lines of development are typical. The first and most obvious is "day care" the professionalizing of childrearing. Professionals, not parents or grandparents, look after even the smallest of infants, is allowing the parents to lead more "productive" lives. In these care centers elements of depersonalization are virtually inevitable. But there is an additional line of development. For economically advantaged families talent development – in contemporary terms "asset management" or "skills enhancement" – is possible, partially motivated by parent guilt, but also as a substitute for hands-on parenting. In this there is much family enrichment, both culturally and financially. Though a child is not yet economically viable, it is nonetheless perceived as a possible asset to the family, and a resource in the making, especially when these early lessons prove to be successful.

Rightly, Hegel and Confucianism see that a significant dimension of who and even what we are is the result of the manner in which we are experienced and treated. The family is central. But in highly competitive, upwardly mobile economic partnerships (marriages) as found in the United States, for example, elements of depersonalization and a corresponding attitude of asset management and development are already at work. Advances in biotechnology and medical science more generally simply extend asset management options and opportunities to an earlier, if not original point in human existence.

As biogenetic engineering becomes more available to the affluent – and becomes more effective and accepted as a means of configuring what infants will become – the empathic sense of self, simultaneously received and conveyed in Confucian terms through *hsin*, is likely to become more at risk (and thus *less* effective) as a means of counterbalancing depersonalizing tendencies. The acceptance of biogenetic engineering starts as the very humane activity of genetically-based disease prevention. It graduates to the enhancement, if not manufacture of various skills predispositions. Probably not to be known until nearly the end of the 22nd century, the final end of this enterprise is quite disturbing. Were a person to be little more than a set of complex skills, including (somewhat paradoxically) person-skills, the nearly literal manufacture of people will become not just theoretically, but practically possible.

Might a class of analog-people arise that are nearly, but not quite people as we have known them? In the advanced technological countries living within an economically driven metaphor and a globally competitive mind-set, this is more than a remote possibility. Once efficiency and productivity become not just *means* but pervasive *endpoints*, it is "not so wild a dream."

First come military uses for such beings, then domestic assignment to chores performed far more efficiently by mistake- and illness-free entities. There are further temptations for the more affluent: the "production" of infant-resources that become in their adult manifestations family assets (or slaves), or superior people (or sophisticated instruments).

In the face of this likely trajectory, families will be needed more and more, yet are also imperiled as centers of sentiment and recognition. Empathy (not itself necessarily sentimental) allows the intimate recognition of people as *persons*. At the least Confucian *hsin* reinforces the experience humans must have of themselves. They must experience themselves as people who have and can provide individual resources, not as resources that happen also and on occasion to be treated genuinely as people.

Biogenetic   engineering   will   further   encourage   tendencies   toward de-personalization. Those economic lenses through which our "advanced," rationalistically driven world more and more sees itself reinforce and are reinforced by these biogenetic opportunities. Clearly, reason is not the best nor the only tool for resisting these temptations. As the most recent manifestation of enlightenment hubris it is currently configured in the service of progress.

But progress toward what? The nourishing of empathy (*hsin*) may not answer this unsettling question. But empathy may help keep us safely on this side of those tipping points I have adumbrated. Genuine family life matters.

VI. CONCLUSION

More immediate questions arise regarding whether and to what extent we have the right to empower infants. One continuing argument brought against our right to engage in truly radical empowerment – something bordering on manufacture or creation – has been that only God has such a right. But we live in multi-ethnic societies with a variety of beliefs, and many within these various societies no longer believe that there is a God, at least not a God of Creation, and some religions require no god at all. Even among those who do hold fairly conventional theistic beliefs, it is thought that the elimination of genetically based predispositions to disease is permissible, and that even the genetic addition or enhancement of certain otherwise absent or enfeebled abilities should best be understood as the appropriate improvement of human life. For these thinkers the difference brought about through genetic research is one of degree, not of kind, from historically longstanding medical and educational practice. Underlying assumptions regarding these matters run so deep and are so subject to conflict that bio-ethicists have little recourse save to science or to religion. Solutions are lived in differing ways within diverse and separate communities of belief and practice. Whether such communities can sustain themselves in viably autonomous ways in the face of the erosion of borders and within a growing and fractured globalizing world is a far more difficult question. Though its answer could not emerge from secular history itself, more sustainable responses to the issues I have raised must clearly await the passage of time.

REFERENCES

Engelhardt, H.T., Jr. (1986). *The Foundations of Bioethics*, (Revised Edition 1996), Oxford University Press, New York.

Engelhardt, H.T., Jr. (2000). *The Foundations of Christian Bioethics*, Swets & Zeitlinger, Lisse, Netherlands.

Engelhardt, H.T., Jr. (2002). 'The Ordination of Bioethicists as Secular Moral Experts,' *Social Philosophy & Policy*, 19, (Summer), 59–82.

Erickson, S. (1999). *The (Coming) Age of Thresholding*, Kluwer Academic Publishers, Dordrecht, Netherlands.

Fan, R. (2002). 'Reconstructionist Confucianism and Bioethics: A Note on Moral Difference,' in H.T. Engelhardt, Jr., and L.M. Rasmussen (Eds.), *Bioethics and Moral Content: National Traditions of Health Care Morality* (pp. 281–287), Kluwer Academic Publishers, Dordrecht, Netherlands.

Fan, R. (2002). 'Reconstructionist Confucianism and Health Care: An Asian Moral Account of Health Care Resource Allocation,' *Journal of Medicine and Philosophy*, 27 (6), 675–684.

Fan, R. (2003). 'Rights or Virtues? Toward a Reconstructionist Confucian Bioethics,' in R. Qiu (Ed.), *Bioethics: Asian Perspectives* (pp. 57–68), Kluwer, Dordrecht.

Fan, R. & Tao, J. (2004). 'Consent to Medical Treatment: The Complex Interplay of Patients, Families, and Physicians,' *Journal of Medicine and Philosophy*, 29 (2), 139–148.

Fan, R. (Ed.) (1999). *Confucian Bioethics*, Kluwer, Dordrecht.

Hegel, G.W. (1977). *Phenomenology of Spirit*, Oxford University Press, Oxford, England.

Qui, R. (Ed.) (2003). *Bioethics: Asian Perspectives*, Kluwer, Dordrecht.

CHAPTER 5

THE MORAL GROUND OF TRUTH TELLING GUIDELINE DEVELOPMENT

The Choice Between Autonomy and Paternalism

SHUH-JEN SHEU

National Yang Ming University, Taiwan

I. INTRODUCTION

The Taiwanese medical system encounters the dilemma of telling or not telling the truth to terminal cancer patients. Most healthcare professionals and the families of these patients believe that it is their ethical responsibility to safeguard vulnerable patients. Protectiveness in the Chinese ethical tradition is profound, whereas, Western health care systems have long considered autonomy to be an important ethical principle in directing clinical decisions. Patient autonomy has been promoted significantly by declaring that it is the moral and legal right of competent individuals to make decisions about the course of their terminal illness. By examining the importance of a patient's wishes and preferences and the context of the family's role in decision-making, the purpose of this article is to search for appropriate ethical reasons for developing a truth telling guideline for terminal cancer cases in Taiwan.

The first part of this paper examines the possible consequences if the health care professionals in Taiwan perceive and carry out their responsibilities when relaying information to vulnerable patients, based on the western principle of autonomy as a guideline for truth telling. The second part of this paper explores the negative impact of truth telling decisions in Taiwan's protective medical system and examines other possibilities. Because a true ethical dilemma goes deeper than a simple "right" or "wrong" answer, arguments based on differing moral grounds are presented. Finally, a clearer picture of the ethical guidelines of truth telling to terminally ill persons is delineated and directions are provided for furthering the understanding of ethics and actions for clinical truth telling in Taiwanese culture.

59

S.C. Lee (ed.), The Family, Medical Decision-Making, and Biotechnology, 59–70.
© 2007 *Springer.*

II. FACING THE DILEMMA OF TRUTH TELLING IN TERMINAL CANCER

The ethics of cancer research has only recently assumed importance in medical literature. Vanderpool and Weiss (1987) have identified 776 articles, books, book chapters, and letters published since 1945, dealing with both ethics and cancer. The recent increase of writing concerning ethics and cancer has become dramatic. Although the majority of articles addressed more than one ethical issue, the most common topic was whether patients should be told the truth about their diagnosis and prognosis (this topic is in 241 of the articles). In most cultures, the final decision on information disclosure lies with the physician. Some medical literature suggests that patients want to be told the truth about a diagnosis of cancer (Aoki, Nakagawa, Hasezawa, Tago, Baba, Toyoda, Toyoda, Kozuka, Kiryu, Iigaki, & Sasaki, 1997; Kashiwagi, 1999; Lin, 1999). Despite this evidence of their patients' wishes, physicians in many countries still hesitate to disclose terminal cancer diagnoses and the related prognosis. It appears that withholding the truth from cancer patients remains a very common issue in Taiwan (Hu, Chiu, Chuang, & Chen, 2002), Hong Kong (Pang, 1999; Tse, Chong, & Fok, 2003), Japan (Akabayashi, Kai, Takemura, & Okazaki, 1999), Greece (Mystakidou, Liossi, Vlachos, & Papadimitriou, 1996; Rigatos, 1997), Italy (Grassi, Giraldi, Messina, Magnani, Valle, & Cartei, 2000), and other countries (Meyza, 1997; Fan, 1997; Uchitomi & Yamawaki, 1997).

According to the literature, the relationship between truth telling and culture has been the subject of increasing attention (Faysman, 2002; Mystakidou et al, 2003; Fan, 1997). The issue of whether, what, how much and how to tell cancer patients concerning their diagnosis, prognosis, or even imminent death are still approached differently depending on various countries and cultures. The dilemma of whether or not a doctor should tell a patient dying of cancer remains a difficult one. Ethical dilemmas are arising from the conflicts of individuals within principles and between individuals holding differing ethical perspectives.

III. GUIDELINE DEVELOPMENT FOR TRUTH TELLING IN AN ASIAN SOCIETY

Developing a standard approach to complex ethical cases in clinical situations allows professionals to incorporate significant problems for an individual while ensuring that appropriate factors are considered.[1] However, it is not easy to establish a standard guideline that will fit in a domestic culture. In 1989, the Japanese Ministry of Health and Welfare and the Japan Medical Association already suggested that physicians should not only inform terminally ill patients of their diagnosis but also of their limited life expectancy. Physicians should first provide the details of the disease to their patients. Thereafter, family members should be informed, but only with the patient's consent. Guidelines for telling the truth to cancer patients formally proposed in 1998 by the National Cancer Center of Japan indicated that the diagnosis must be discussed first with the patients themselves whenever possible. However, giving the diagnosis of cancer to patients remains controversial and the trend towards full disclosure of the diagnosis or the prognosis has progressed only slowly (Seo et al., 2000).

Despite the increasing concerns of truth disclosure, most studies reflected that difficulties about telling the truth to patients still existed in Japan (Akabayashi et al., 1999; Horikawa, Yamazaki, Sagawa, & Nagata, 1999; Mizuno, Onishi, & Ouishi, 2002; Morioka, 1991; Seo et al., 2000; Swinbanks, 1989; Tanida, 1994). Several authors have tried to provide some insight into this issue by evaluating results from questionnaires given to hospital patients, families, or doctors in a mass survey. Most results showed that physicians wanted to tell or the patients intended to be told the truth, but only few patients responded that the truth was actually told. It is also reported that all family members were told about the patient's condition in detail, while few patients were informed of the diagnosis of cancer. It is still a Japanese custom that physicians usually explain the patient's condition to family members first and that the cancer patients are informed only when their families agree on what to tell them. This common Japanese practice appears to be similar to the Taiwanese culture of family collectivism, which indicates that the sense of belonging to the family is very strong.

It is now time to reconsider and leave behind the common Japanese or Taiwanese interpretation of truth telling or informed consent – for example, respecting the wishes of family members rather than those of the patient. It also must be examined whether it is appropriate to advocate that all information including the present condition, diagnosis, diagnostic procedures, and treatment plan should first be told to the patient by an experienced physician and thereafter, the family members and acquaintances that the patient wishes to be informed can be given the opportunity to hear the physician's explanation about the patient's condition. The official Japanese medical value system and governmental guideline favored giving priority to patient autonomy. However, other people may feel that rather than bureaucratic policies, the doctor's duty of beneficence or a certain indigenous culture can be the overriding principle governing such decisions. On the other hand, physicians frequently ignore their patients' wishes when they consider the appropriateness of truth telling. It seems that both approaches of governmental paternalism or professional autonomy have something to offer. However, a complete shift from primarily family disclosure to mandatory disclosure to patients without considering sophisticated communication skills and patients' or families' preferences and readiness may result in serious harm to them.

IV. DEALING WITH AUTONOMY, PATERNALISM, OR VULNERABILITY OF TRUTH TELLING IN TAIWAN

Will the principles of autonomy upheld by the Western bioethics meet the needs and challenges in Asia? Is it appropriate to just adopt the dominant western concept of patient autonomy into the Asian form of truth telling? Within a clinical context, there is traditionally a disparity between patient autonomy and family or professional paternalism. Both lying and truth telling carry risks of harm to patients. The "overriding" nature of a family, physician, or governmental paternalism surpasses the patient's autonomous preferences, decisions, or actions out of a concern for the patient's welfare. Conversely, the principle of patient autonomy necessitates the empowering of patients through the provision of information. However, in connection with the choices that patients wish to make about controlling the end of their lives, families and doctors may also find only the emphasis on the patient's autonomy to be problematic. Learning to work with and how to balance these risks can be an important goal of medical ethics guideline in the Asian society.

One study examined a group of 195 Taiwanese subjects at a medical university hospital regarding their attitudes toward truth telling of cancer. It is found that once diagnosed with cancer, 92.3% of the participants preferred being told the truth about their diagnosis and 7.7% did not. A total of 62.6% of the participants preferred that doctors tell a relative the truth about their cancer diagnosis, while 37.4% preferred that doctors not tell a relative the truth. The authors concluded that a majority of patient subjects in Taiwan would prefer to know the truth if victimized by a cancer disease, despite the supposed influence of Chinese culture. Furthermore, attitudes toward the truth telling of cancer differed between relatives of patients and the patients themselves. Relatives of cancer patients were more likely to follow to the principle of beneficence, whereas the patients themselves were more likely to follow the principle of autonomy (Wang, Chen, Chen, & Huang, 2004). Another study posited that the most severe obstacle of the truth telling in Taiwanese clinics is often the family (Hu et al., 2002). The majority of physicians tell the truth more often today than in the past, but most of them prefer to disclose the truth to the next of kin, especially in Asian countries. Reluctance to directly share the truth with the patient about his or her diagnosis and prognosis is frequently associated with cultural pressures and clinical reality.

Within the Taiwanese clinical context, there are three parties existing among patient autonomy, family and professional paternalism. In most cases, patients' families exclude the patient from the process of information exchange in Chinese culture in an effort to protect them from despair and a feeling of hopelessness (Tse et al., 2003). The apparent major problem of truth telling in Taiwan is that an ill patient often loses a chance of self-determination, and he or she might be thought of as an incompetent person, who is in need of protection. Truth telling to a perceived vulnerable person is not completely accepted by the majority of people. Patients who are confronted with the autonomy and paternalism dichotomy, are often unable to verbalize their needs or wants to their families and doctors, either because they are intimidated by the doctors' perceived power or the doctors lack the time and the techniques to listen to their concerns, or because their families are emotionally hurt, but still pretend that they should and can protect the patients. It is not difficult to acknowledge the vulnerability both from the view of the patient and her or his family.

The disclosure of diagnosis of terminal cancer may affect the individual and family structure and interpersonal dynamics in different perspectives and extents. In fact, most health professionals and families are ambivalent about telling the truth to patients. Disclosing the truth would become an insincere act if a patient were to lose hope and confidence in life after learning of his or her disease. In this system of protective medical care, it is arguable as to whose interests are being protected: the patient's, the families' or the professionals'. Some authors suggest that honoring the interests of the hospital and family members who legitimately represent the patient's interests may be at the expense of the patient's right to know (Pang, 1999). There are some cases in clinics where the physician did not disclose the truth to the patient, because of fear of recrimination and litigation by the family. The act of suing usually implies that the family shifts their emotional and moral frustrated reactions toward the physician. The family tries to resist their own vulnerable feelings and keep balanced by using this limited power to induce and control the physician's vulnerability.

V. FIGURING OUT AND REINTERPRETING THE MEANING OF AUTONOMY AND PATERNALISM

The apparent disparity between the 'patient knows best', the 'family knows best', or the 'doctor knows best' arguments forms the multiple autonomy and paternalism dichotomy. It should not be doubted that patients are in the foremost position to decide on whether or not and what they are worth achieving. Correspondingly, they are in a prime position to evaluate, interpret and contemplate available medical information and to make evaluative judgments about what modality is appropriate and worth trying. This is generally accepted as true because only patients know how much risk they are willing to take and can judge particular benefits. However, the outcome of medical interventions does not affect patients alone.

The cultural life of a community might determine how individuals and families are valued and what kind of measures physicians and families can take to disclose the truth of illness. After a survey of 800 seniors from four different ethnic groups, Blackhall (2001) showed that European-American and African-American respondents tend to believe that a patient should be told the truth about the diagnosis and prognosis of a terminal illness and they were more likely to view truth telling as empowering and enabling the patient to make choices. On the contrary, the Korean-American and Mexican-American respondents were unwilling to hear the bad news and were more likely to see truth telling as cruel and even harmful to the patients. In Blackhall's previous article, it was concluded that Korean-American and Mexican-American subjects were more likely to hold a family-centered model of medical decision-making rather than the patient autonomy model favored by most of the African-American and European-American subjects. The authors suggested that physicians should ask their patients if they wish to receive information and make decisions or if they prefer that their families handle such matters (Blackhall et al., 1995). For solving family-related barriers to truthfulness in Taiwanese cases of terminal cancer, Hu, Chiu, Chuang, and Chen (2002) suggested that health professionals could communicate with families first and discuss the possible emotional reactions from patients, give patients enough time to reflect on their sicknesses and discuss further what patients have been told, and then disclose information based on patients' expectations and support them.

The general view is that Asian people embrace a more family-oriented and less individualistic approach to truth telling than Anglos. In the family centered mode, it can become the sole responsibility of the family to hear bad news about the patient's terminal illness and to make difficult decisions about life support. Although the patients are consequently best situated to recognize who should be significantly involved in making determinations, it seldom occurs because patients are often considered to be insufficiently authoritative or incompetent to make judgments. To give vulnerable patients the power of right to know or not to know or forcing them to make decisions abruptly would be uneasy for them too. Nevertheless, people may be adamant in retaining control of their end-of-life care decisions while they are capable of doing so, and they may also want their proxy of their choice to retain control of doing so if they became incapable to do so.

Tse, Chong, and Fok (2003) argue that the Chinese views on autonomy and nonmaleficence do not really justify non-disclosure of the truth. They also recommended that truth telling should mainly depend on what the patient wants to know and is prepared to know, and not only on what the family wants to disclose. The authors classified the conditions of family determination into three levels: 1) The family takes part in decision making with the patient, 2) The patient asks the family to decide, 3) The family decides alone despite the patient's wish to participate. Only the third level represents strong paternalistic style and excludes the patient and is against patient autonomy in the broad sense. Justification of this paternalistic approach depends on the principle of nonmaleficence. It cannot be justified solely by the concepts of harmonious dependence or relational self.

In postmodern medicine, emphasis on patient autonomy is still important and it can change not only the nature of the doctor-patient relationship (Moodley, 2003), but also the family-patient relationship. On the other hand, Pang (1999) posited that establishing a paternalistic medical or family attitude in the relationship also means the capacity and competence to meet the needs of patients. Keeping an appropriate level protectiveness (but not overprotecting or cheating the patient) is important. We should not only view truth telling to the terminally ill as being primarily governed by the principles of paternal reality or respect for autonomy. A single sided approach may only reveal the superlative quality and conflict between nonmaleficence and beneficence and autonomy with regard to truth telling having originated from a narrow understanding of the concept of autonomy or paternalism. All such rules are of limited value in medical ethics. We should instead turn to an ethics deriving from the centrality of moral relationships and virtues that would reduce the value that philosophical fashion places on truth telling. The terminally ill truth telling should be more dependent on the circumstances of particular cases (Byrne, 1990; Taboada & Bruera, 2001).

Much of the contemporary literature in the field of bioethics emphasizes the importance of truth telling, the principle of autonomy and the significance of patient choice, the value of advance care planning, and the right to have treatment withdrawn or withheld. However, a growing body of scholarship on the cultural dimensions of end-of-life care reveals a plurality of attitudes toward what constitutes appropriate forms of communication and decision making. While there are limits to the kind of social practices and moral claims that should be accommodated in liberal democracies, health care providers should take steps to ensure that they are providing culturally sensitive end-of-life care (Turner, 2003).

VI. A CULTURAL SENSITIVITY AND TEAM WORK APPROACH TO TRUTH TELLING

Culture shapes what actions and kinds of behavior are considered acceptable and serves as a guide to persons on the ways in which they should fulfill the needs of others. Turner (2003) suggested that scholarship in bioethics need to better address complex ethical issues in settings where many cultural modes and religious traditions can be found, and where disagreement frequently arises about basic questions of reasonableness, professional standards, and family obligations. While shared standards and norms exist, the study of ethical issues in end-of-life care reveals the existence of different social practices and models of moral deliberation. Tanida (1994) indicated that the Japanese attitude toward avoiding truth disclosure stems primarily from paternalism but is also influenced by social characteristics including the insufficient understanding of this issue. Open discussion involving all factions of society is necessary to attain a better understanding of this issue and to promote eventual truth disclosure. Health care professionals should be sensitive to culturally defined health beliefs and practices that may explain the behavior of individuals or families such as the degree and quality of parents' or families' readiness and involvement in patient care, the patient's relationship with the family and health care staff, and its adherence of meaning and possible ways of truth telling. For example, some elders in Taiwan would like to acknowledge their failing body conditions as a part of the aging process rather than admitting it as the cause of cancer. Some might think the cancer as a curse from a supernatural myth while others might not mind talking about the death issue at all. In many cultures, individual autonomy is not the central core of identity. Instead, persons are seen as embedded in the family and community within a complex web of obligations that create interdependence (Candib, 2002) .

The professional stipulation of autonomous individual comes into conflict with the family's moral constraint – that is, parental or filial duty. In some cultures, when one of the family members gets ill, the important relatives will carry on the protective responsibility for the ill one. However, within the same ethnicity, there is the effect of obscuring the heterogeneity of the groups, such as differences in terms of belief, language, religions, cultural customs, and relationships. It is unlikely to assemble uniformly around universal values and the culture evolves over time. For example, in older generations, the young ones might not be particularly interested in exercising their right to make decisions and they respect their elders' opinions more. They prefer to use the principle of benevolence and embrace a virtue-oriented approach to ethical issues, to a certain extent, that gives priority to community and family relationships over their personal right to make their own choices. Conversely, the modern younger generations are profoundly focused on their own rights and principles. Following the principle of autonomy, it might help counterbalance the paternalism prevalent in some current cultural or medical practices.

By questioning the conception that links autonomy with the choice of knowing the truth or not, some feminist and other scholars' responses to individualism challenge the assumption that increased opportunities to make choices in clinics will either enhance autonomy or family relationships. Scholars argue that the crux of the problem arises from the liberal individualistic understanding of autonomy and human efficacy in terms of choice, rights, and self-determination. It is inappropriate to see other people as separate and autonomous agents but rather it is important to see them involved in a unique relationship and in a specific community (Mackenzie, 1984 & 2000). Pratt (1991) pointed out that the cultural tendency of emphasizing self-actualization, self-reliance, and "finding yourself" in the United States often encourages Americans to develop a sense of self-sufficiency and personal autonomy that can have the unintended effect of distancing them from their original family ties. In this conceptual system, this distancing results in a search for other forms of intimacy and emotional attachment as a replacement for the family. The relationship with family members is very much a voluntary one, subject to the vicissitudes of a mobile society that also relies on government and social institutions for the provision of welfare, health, and eventual care during the later life.

While the principle of respect for autonomy is sometimes seen as a Western, ethnocentric notion, there is considerable emphasis upon individual choice in bioethics scholarship and it also can be viewed as a practical mechanism for respecting cultural and religious differences. Indeed, it can also be appropriate to consider plural traditions of moral reasoning and there is good reason to think that a plurality of human religious and cultural practices should be accommodated. However, this can pose a considerable challenge for bioethicists, and clinicians because it can lead to difficulties regarding "closure" or "disclosure" in moral reasoning (Turner, 1998). Within multicultural settings such as Canada and the USA, respect for cultural and religious diversity is an important element of political, personal, professional, and institutional life. Some cultural and religious practices are not morally or legally acceptable because these actions risk seriously harming others and undermining basic human rights. There are limits to tolerance and accommodation in multi-ethnic, religiously diverse societies. There are many reasons for promoting respect for a wide range of substantive moral norms and social practices (Turner, 2003). The dangers in addressing the moral obligations of healthcare providers in multicultural, pluralistic settings are those of falling into a simplistic acceptance of all cultural and religious norms, even when some practices cause great harm and violate basic human rights, and insisting upon a narrow understanding of acceptable moral reasoning.

The complexity of the truth-telling task is more acute in end-of-life decision making, when decisions must be made in a timely manner, under highly emotional circumstances, without the benefit of retrospection (Turner, 2003). Sherwin (1998) posited that individuals tend to be treated as interchangeable in that no attention paid on the details of personal experience although Bok (1978) reminded us of the possible negative consequences of lying to patients. The improved understanding and better knowledge about how to manage terminally ill patients can alleviate anxiety among patients, families, and professionals. The standard palliative care approach to breaking bad news should be adopted, but with some modifications to address the family determination (Tse et al., 2003) or the individual autonomy. Assessment of health beliefs and practices, family composition, religious rites, and beliefs about illness causation could be routinely made on each family (De & Kovalcik, 1997). Clinical and bioethics professionals should also be aware of how their own cultural and social backgrounds influence the way they perceive ethical dilemmas and remember to make room for the diverse views of the populations they serve (Blackhall, 2001). Further directions should be noted in how the truth should be told and even in definitions of what constitutes "truth" and "telling". The health care team, patient, and family relationship is a triangle where each part supports the other two and is affected by the cultural background of each of the others as well as the changes that occur within the triangle (Mystakidou et al., 2003). A multi-disciplinary and team work approach should be available to be adopted in designing a guideline for truth telling. The assignment of multidisciplinary commission can contribute to both the patient's and family's comfort in different perspectives. It can also help the physician to better make the decisions and compensate for negative consequences.

Ideally, health professionals need to involve and enquire more actively about cancer patients' concerns and feelings. Narrative ethics can become a useful concept and tool for gaining insight into the care of hospitalized dying patients. Listening on the individual's narratives can help script more empathic and compassionate care of the dying, and grasp the relationships between patients, families and clinicians; and the influence of time, uncertainty, ambiguity, resource allocation, and spirituality on the course of care (Fins, Schwager, & Acres, 2000). They also need more training in the relevant interviewing skills to conquer their fear that truth telling will damage patients psychologically. In considering how their interviewing skills might be improved, the key question is which interviewing behaviors promote patient disclosure and which inhibit it. There are some other suggestions about the utility of specific interviewing behaviors, such as recognizing the patient's identity, humor, the effect of the aesthetic and spiritual environment, being watched or chosen isolation, relinquishment and relaxation, being safe and keeping control, awareness of mortality, and recognition of the palliative care philosophy (McKinlay, 2001). Focusing on dialogue, deep listening and reflection in action can be critical in opening communications paths between the dying patient and his or her families (Cherin, Enguidanos, & Brumley, 2001).

VII. CONCLUSION

Patients, families, and professionals can inhabit distinctive social worlds where they are guided by diverse understandings of moral practice. Decisions made by the family or the physician may not always reflect the patient's wishes. Denying information disclosure or telling the truth to the patient is probably more a historical, emotional issue rather than an ethical, cultural phenomenon. There are intergenerational differences and conflicts between the very conservative and the democratic ways to tell the truth in domestic culture. People are committed to preserving or disclosing the truth in various extents and contexts. Health professionals need to realize not only the right of patients to know or not to know but also they need to have more capability to know their clients and recognize them as a unique social, emotional, and humanistic body with certain cultural and language conditions. Greater understanding and reconstructing the meaning of autonomy and paternalism allow us reason and act morally better. Further, multidisciplinary teamwork interaction and communication may guide illness disclosure activity more successfully. It is important to decrease the risk of harm and increase the value of human life by clear moral, situational thinking and action. By being sensitive to what and how much the patient wishes to know and by learning from the insights provided by various personnel's points of view, policy makers and clinicians may learn how to make better moral decisions and take action in these areas.

NOTE

[1] The original field study related to this article was funded by the Bureau of Health Promotion, Department of Health, Taiwan, R.O.C., a grant (DOH 92-HP-1507), named "A triangulated approach of exploring the phenomenon of truth-telling cancer medicine from the perspectives of health professionals, patients, and families". One of the most important goals of this study was to develop a truth telling guideline for the terminal cancer patients in clinics. This article serves to explicate and analyze the important ethical concepts in order to establish a sound moral ground as a basis for truth telling guideline development in Taiwan.

REFERENCES

Akabayashi, A., Kai, I., Takemura, H., & Okazaki, H. (1999). 'Truth telling in the case of a pessimistic diagnosis in Japan,' *Lancet*, 354, 1263.

Aoki, H., Molkentin J.D., Cowley, A.W., Jr., Izumo, S., Markham, B.E., Herzig T.C., & Jobe, S.M. (1997). 'Significance of informed consent and truth-telling for quality of life in terminal cancer patients,' *Radiation Medicine*, 15,133–135.

Blackhall, L., Murphy, S., Frank, G., Michel, V., & Azen, S. (1995). 'Ethnicity and attitudes toward patient autonomy,' *JAMA*, 274, 820–825.

Blackhall, L., Frank, G., Murphy, S., & Michel, V. (2001). 'Bioethics in a different tongue: the case of truth-telling,'*Journal of Urban Health*, 78, 59–71.

Bok, S. (1999). *Lying: Moral Choice in Public and Private Life: Lies to the Sick and Dying*, 2nd Edition, Vintage Books, 220–241.

Byrne, P. (1990). 'Comments on an obstructed death – a case conference revisited: commentary 1,' *Journal of Medical Ethics*, 16, 88–89.

Candib, L. (2002). 'Truth Telling and Advance Planning at the End of Life: Problems with Autonomy in a Multicultural World,' *Families, Systems & Health: The Journal of Collaborative Family Health care*, 20, 213–229.

Cherin D., Enguidanos S., & Brumley R. (2001). 'Reflection in action in caring for the dying: applying organizational learning theory to improve communications in terminal care,' *Home Health Care Services Quarterly*, 19(4), 65–78.

De, T. & Kovalcik, R. (1997). 'The child with cancer. Influence of culture on truth-telling and patient care,' *Annals of the New York Academy of Sciences*, 809, 197–210.

Fan, R. (1997). 'Truth telling to the parent: cultural diversity and the East Asian perspective,' in N. Fujiki and D. Macer (Eds.), *Bioethics in Asia* (pp. 127–129), Eubios Ethics Institute.

Faysman, K. (2002). 'Cultural dimensions of anxiety and truth telling,' *Oncology Nursing Forum*, 29, 757–759.

Fins, J., Schwager, G., & Acres, C. (2000). 'Gaining insight into the care of hospitalized dying patients: an interpretative narrative analysis,' *Journal of Pain and Symptom Management*, 20, 399–407.

Grassi, L., Giraldi, T., Messina, E., Magnani, K., Valle, E., & Cartei, G. (2000). 'Physicians' attitudes to and problems with truth-telling to cancer patients,' *Supportive Care of Cancer*, 8, 40–45.

Horikawa, N., Yamazaki, T., Sagawa, M., & Nagata, T. (1999). 'The disclosure of information to cancer patients and its relationship to their mental state in a consultation-liaison psychiatry setting in Japan,' *General Hospital Psychiatry*, 21, 368–373.

Hu, W., Chiu, T., Chuang, R., & Chen, C. (2002). 'Solving family-related barriers to truthfulness in cases of terminal cancer in Taiwan. A professional perspective,' *Cancer Nursing*, 25, 486–492.

Kashiwagi, T. (1999). 'Truth telling and palliative medicine,' *Internal Medicine*, 38, 190–192.

Lin, C. (1999). 'Disclosure of the cancer diagnosis as it relates to the quality of pain management among patients with cancer pain in Taiwan,' *Journal of Pain and Symptom Management*, 18, 331–337.

Mackenzie C. & Stoljar, N. (2000). *Relational Autonomy: Feminist Perspectives on Autonomy, Agency and the Social Self*, Oxford University Press, New York.

Mackenzie C. (1984). *Caring: A Feminist Approach to Ethics and Moral Education*, University of Berkeley, Berkeley.

McKinlay, E. (2001). 'Within the Circle of Care: patient experiences of receiving palliative care,' *Journal of Hospice & Palliative Care*, 17, 22–29.

Meyza, J. (1997). 'Truth-telling, information, and communication with cancer patients in Poland,' *Annals of the New York Academy of Sciences*, 809, 468–479.

Mizuno, M., Onishi, C., & Ouishi, F. (2002). 'Truth disclosure of cancer diagnoses and its influence on bereaved Japanese families,' *Cancer Nursing*, 25, 396–403.

Moodley, K. (2003). 'Respect for patient autonomy,' *The Journal of the Dental Association of South Africa*, 58, 323.

Morioka, Y. (1991). 'Informed consent and truth telling to cancer patients,' *Gastroenterologia Japonica*, 26, 789–792.

Mystakidou, K., Liossi, C., Vlachos, L., & Papadimitriou, J. (1996). 'Disclosure of diagnostic information to cancer patients in Greece,' *Palliative Medicine*, 10, 195–200.

Mystakidou, K., Parpa, E., Tsilika, E., Katsouda, E., & Vlahos, L. (2003). 'Cancer information disclosure in different cultural contexts,' *Supportive Care of Cancer*.

Noddings, N. (1986). *Caring: A Feminist Approach to Ethics and Moral Education*, University of California Press.

Pang, M. (1999). 'Protective truthfulness: The Chinese way of safeguarding patients in informed treatment decisions,' *Journal of Medical Ethics*, 25, 247–254.

Pratt, D.D. (1991). 'Conception of self within China and the United States: Contrasting foundations for adult education,' *International Journal of Intercultural Relations*, 15, 285–310.

Rigatos, G. (1997). 'Cancer and truth-telling in Greece. Historical, statistical, and clinical data,' *Annual New York Academy of Sciences*, 809, 382–392.

Seo, M., Tamura, K., Shijo, H., Morioka, E., Ikegame, C., & Hirasako, K. (2000). 'Telling the diagnosis to cancer patients in Japan: attitude and perception of patients, physicians and nurses,' *Palliative Medicine*, 14, 105–110.

Sherwin, S. (1998). 'A Relational Approach to Autonomy in Health care,' in The Feminist Health Care Ethics Research Network (Ed.), *The Politics of Women's Health: Exploring Agency and Autonomy* (pp. 19–47), Temple University Press, Philadelphia.

Swinbanks, D. (1989). 'Medical ethics: Japanese doctors keep quiet,' *Nature*, 339: 409.

Taboada, P. & Bruera, E. (2001). 'Ethical decision-making on communication in palliative cancer care: a personalist approach,' *Supportive Care of Cancer*, 9, 335–343.

Tanida, N. (1994). 'Japanese attitudes towards truth disclosure in cancer,' *Scandinavian Journal of Social Medicine*, 22, 50–57.

Tse, C., Chong, A., & Fok, S. (2003). 'Breaking bad news: a Chinese perspective,' *Palliative Medicine*, 17, 339–343.

Turner, L. (1998). 'An anthropological exploration of contemporary bioethics: the varieties of common sense,' *Journal of Medical Ethics*, 24, 127–133.

Turner, L. (2002). 'Bioethics and end-of-life care in multi-ethnic settings: cultural diversity in Canada and the USA,' *Mortality*, November, 7, 285–302.

Uchitomi, Y. & Yamawaki, S. (1997). 'Truth-telling practice in cancer care in Japan,' *Annals of the New York Academy of Sciences*, 809, 290–299.

Vanderpool, H. & Weiss, G. (1987). 'Ethics and cancer: a survey of the literature,' *Southern Medical Journal*, 80, 500–506.

Wang, S., Chen, C., Chen, Y., & Huang, H. (2004). 'The attitude toward truth telling of cancer in Taiwan,' *Journal of Psychosomatic Research*, 57, 53–58.

CHAPTER 6

TRUTH TELLING TO THE SICK AND DYING IN A TRADITIONAL CHINESE CULTURE

STEPHEN WEAR

The University at Buffalo, Buffalo, New York, United States

Traditionally and currently, Chinese medicine has generally not embraced the notion that the sick and dying should be told the truth about their situation and prospects, opting instead for conveying this information to family members who assist practitioners in making medical decisions (Chan, 2004; Cong, 2004; Fan and Benfu, 2004). Interestingly, in Taiwan specifically, informed consent, by law, can be secured from either the patient, or a family member, with (I am informed by local authorities) a vast preference for the latter approach by most parties.

Western bioethics appears to have an ambivalent attitude about such a practice. On the one hand, extensive discussions in the West over the past three decades have generally enshrined truth telling as a basic ethical principle with strong and detailed argument in support of this primacy. (Annas, 1975; Parsons, 1975; Ramsey, 1970) At most, in the West, current discussions regarding truth telling relate mainly to possible exceptions to the generally held rule, and are seen by many to be quite difficult to establish (Meisel, 1979; Wear, 2004). This is not, furthermore, merely a scholarly consensus. At least in the USA, the primacy of truth telling is enshrined in prominently displayed patients' rights statements, institutional policies, and in the standards of hospital accreditation organizations, e.g. the Joint Commission for the Accreditation of Hospitals (JCAHO). Simply, it is uncontroversial and expected. (Wear, 1993)

On the other hand, western bioethics, and ethics generally, certainly comes with the basic caveat that "ethical principles" such as truth telling, are culturally conditioned, and may well have meaning and force only in a culture that shares basic western assumptions regarding personal autonomy, independence, and the subjectivity of the values. If we move to a different culture, e.g. traditional Chinese culture, that emphasizes contrary values such as seeing families as the decision making unit, and the objectivity of values, then truth telling may well have no special status, and its opposite, viz. lying to, deceiving or not being candid with patients, may well be the guiding ethical principle (Fan and Benfu, 2004).

71

S.C. Lee (ed.), The Family, Medical Decision-Making, and Biotechnology, 71–82.
© 2007 *Springer.*

This paper will challenge this culturally accommodating view in a very specific sense. If it were the case that Western bioethics were merely just another world view, based primarily on a deontological or rights based preference for truth telling in the service of autonomy, then it would arguably have little to say to the East. But this is not the case. The Western preference for truth telling is also bolstered by the clinical experience of the specific harms and benefits of truth telling to patients, and its opposite. In sum, the experience in the west has strongly and increasingly suggested that the consequences of the practice of truth telling are markedly superior to the practice of lying and deception. And whatever ethical theory or view one has, one should wonder if it would be worthy of the name if it is oblivious to consequences.

However, this paper will studiously avoid presuming to take a firm stand on whether truth-telling to the sick and dying should also become the rule in traditional Chinese societies. This is so as the "data is not in" on at least two crucial points that go to the heart of the potential effectiveness and appropriateness of truth telling in such societies. For one thing, it is not at all clear what Chinese patients typically want. Only rudimentary studies seem to be presently available in this regard, and their results are contradictory, i.e. some find a marked preference for truth telling among Chinese persons (Fielding, 1996), others find the opposite preference (Blackhall, 1995). Until we know that whether an intervention is likely or unlikely to be well received by its recipients, we should maintain a cautious view of its appropriateness. Secondly, the possibility that truth telling to the sick and dying might undermine the traditional Chinese family is surely at issue here. Especially given the fragmenting, destructive effects of post-modern industrial society, we should be loathe to institute something that might weaken support systems that are often crucial to peoples' well being, as patients or otherwise. In effect, for all this paper proceeds by retailing a consequentialist view that strongly favors the practice of truth telling, it will remain quite tentative as to whether such a practice should be inserted into Chinese society.

With these caveats in the background, this paper will proceed, nonetheless, by arraying the Western analysis of the consequences of the opposing practices of truth telling versus lying to or deceiving patients, an analysis that strongly favors truth telling. This will be attempted by first reviewing a case that was once occurred in Taiwan, a case that appears to amount to an obvious example of when traditional Chinese practice is simply unethical. We will then review the arguments of Sissela Bok, in her classic (for western bioethics at least) article "Lies to the Sick and Dying," which retails the core of the consequentialist arguments in this area (Bok, 1978, pp. 232–255). This paper will then move on to reflect on what such arguments may entail for traditional Chinese medicine; in effect, without all the data in, as previously noted, what tentative suggestions might be made about Chinese practice. To anticipate, this paper will presume to suggest that these consequentialist arguments are weighty enough to call for a significant revision to the customary approach of traditional Chinese medicine regarding truth telling, but that this revision will need to be accomplished with a strong sensitivity to the traditional role of the Chinese family. A simple way to put this final recommendation is that the goal of truth telling merits a much more significant place in Chinese medical practice, but that the means by which this place is to be secured will need to key, in many important respects, to the institution of the Chinese family.

I. CASE SUMMARY

Miss G. is a 36 year old female from southern Taiwan who works in a textile factory in Chung Li. On June 20th she presented to a local health clinic with complaints of feeling ill for a couple months as well as having had a 10 kg. weight loss during this period.

As part of the patient's workup, an x-ray of Miss G's chest was done. On review, it was found to show a density in the right upper lobe of her lung. Her doctor recommended to her that she receive broncoscopy to further evaluate what the physician described to the patient as a "shadow" on her lung x-ray. The patient agreed and the broncoscopy was performed.

While awaiting the results of the broncoscopy, Miss G remained in hospital for further treatment and was joined by various family members, including her mother and two brothers. The patient gradually regained weight and felt much better over a two week period.

Subsequently, the results of the broncoscopy documented adenocarcinoma of the lung. The family members were then approached by the clinical staff with a strong recommendation for surgical resection of this tumor, followed by adjuvant chemotherapy and radiation. Staff strongly felt that the tumor was the primary and had not yet metastasized elsewhere. They felt the chance of complete cure was 80%+. Without this intervention, staff was sure the tumor would quickly metastasize and become untreatable and lethal.

The family completely rejected this course of action. They reported a complete faith in the power of herbal medicine and intended to take Miss G. home to southern Taiwan for such treatment. They also rejected the staff's request to discuss any of this with the patient. They demand that the patient simply be discharged to their care.

The families' wishes were honored. About 4 months later, the patient returned to the hospital very ill, metastatic and quite untreatable.

Now whether the ethical evaluation of any case should be seen as intuitively obvious is an issue for ethical theory and wide ranging reflection. But this case would appear to be a good candidate for the ethically obvious, if any case is. That is: whatever else might be said about the case, this young lady ended up dead because her family rejected probably curative surgery AND did not give her the chance to decide otherwise.

Avoiding ethical intuitionism, however, the credibility and usefulness of which seems to evaporate very quickly as soon as we approach less obvious cases, we might at least presume to say this case violates the cardinal principle of beneficence. But even this is arguable as the overall non-candid practice of traditional

Chinese medicine might still be beneficial for patients in some summary sense, and this case only represents the expected tragic exception to an otherwise meritorious practice. One's initial ethical intuition about the case at hand might, in effect, be overwhelmed by broader intuitions and experience regarding which practice truly produces the "greatest good for the greatest number" or some such calculation.

Or one might argue that truth telling should occur as an exception when the best interests of a specific patient are clearly being violated or ignored, as in the case above. But this move only seems to initiate the issue, not resolve it, as there would certainly be much discussion about what any given patient's best interests are, including the autonomy interests that the West seems so concerned to emphasize and protect as opposed to the familial concerns that the East is at least equally concerned with.

The approach of this paper will be to take this case as merely suggesting that traditional Chinese practice may harm some patients and thus merits review. This review, however, will be pursued on a systemic, not a case by case basis. That is: what are the pros and cons, the good and bad consequences, of the alternative practices of lying to or deceiving patients, or telling them the truth?

II. SISSELA BOK'S CLASSIC WORK ON LIES TO THE SICK AND DYING

There is perhaps no more robust an example of a pivotal, watershed contribution to a bioethics issue in the West than Sissela Bok's chapter on "Lies to the Sick and Dying" in her book *Lying: Moral Choice in Public and Private Life* (Bok, 1978, pp. 232–255). Often reprinted in introductory bioethics texts, it might well be the most often read bioethics article in the world, especially among the now millions of undergraduates, medical students, nurses and residents who have taken some form of bioethics courses in the West over the past couple of decades. For the sake of this paper, the approach and arguments of this article will now be summarized.

Bok's article can be divided roughly into two sorts of reflections, the first concerning the pros and cons of the practice of lying to patients, the second regarding the contrary practice of truth telling. Her approach would seem to be best summarized as what we call "rule-utilitarian". In sum, one seeks a guiding rule of behavior, that however much exceptions to the rule might be countenanced, the rule states what one should generally be doing, exceptional circumstances notwithstanding. And the argument is relentlessly consequentialist, i.e. what are the negative and positive consequences of the rule, and its opposite, and which approach is likely to have the more favorable benefit to harm ratio in the last analysis. We should first describe, via Bok and others, what were the basic sorts of arguments in support of the practice of lying to, deceiving or not being candid to patients, a practice that was as dominant in the West a few decades ago as it presently is in the East.

2.1 Arguments in Support of the Practice of Lying to, Deceiving or Not Being Candid with Patients

Western medicine, over the past 2,500 years, has shown very little concern for or commitment to patient autonomy (Katz, 1984). Aside from an occasional rebel, the overwhelmingly dominant ethical orientation of Western medicine has keyed to what will benefit, or at least not harm patients. And the calculation, up until recently, has consistently been that truth telling is likely to do more harm than good, and thus should occur sparingly, if at all. (Ravitch, 1978) The guiding Western ethical principles have thus been non-malificence and beneficence, i.e. not to harm and to benefit, and truth telling has been historically seen as often violating both of them.

Contrary to some who saw all this as just some insidious power trip by doctors, we should proceed by taking such a practice at face value as the conscientious view of people committed to patient's best interests, and who saw truth telling as not among these. The arguments for such a paternalistic approach may be divided into two parts: (1) the harms of truth-telling, and (2) the benefits of lying to, deceiving or not being candid to patients.

(1) The Harms of Truth Telling: most basically perhaps, Western medicine has traditionally seen truth telling to patients as attempting to offer information to people who neither wanted it, nor were capable of using it correctly if provided with it. The common view was that patients do not want the truth, and do not see themselves as decision makers in need of it. Moreover, patients were seen as generally not capable of being informed and understanding medical scenarios and choices accurately, and as likely to make bad decisions, decisions that flow from untutored, erroneous and idiosyncratic views of their situations and prospects. (Ravitch, 1978) Much was concurrently made of the presence of "diminished competence" in patients, i.e. that sick folks are quite likely to be burdened by factors that directly undermine their abilities to understand and make decisions, e.g. fear, pain, stress, the effects of the disease, such as hypoxia, the effects of the treatment, such as pain medications. (Wear, 1993) One might be reminded of the old saying that one should not attempt to teach gorillas to dance because they cannot do it... and it only annoys the gorillas.

Aside from this, numerous substantial harms were predicted, including that truth telling would only depress, confuse and distress patients in general and, when bad news was conveyed, tend to deprive patients of hope and destroy the quality of their remaining time. Equally, patients might withdraw from potentially beneficial treatment and occasionally even commit suicide. In such circumstances, it was postulated, truth telling might also trigger a death mechanism where the loss of hope leads the patient to give up entirely, even biochemically. And all this was suggested within the context of a robust recognition that medical diagnoses and prognoses are quite uncertain, and thus these harms might occur from the provision of false information masquerading as the truth.

(2) The Benefits of Lying to, Deceiving or Not Being Candid to Patients: avoiding truth telling, the benefits of the contrary practice were held to be substantial and

include keeping the appropriate decision makers, e.g. physicians, with some input from patients' families, in their proper role. Similarly, it was commonly held that optimistic patients do better therapeutically, and such optimism was felt to be an important focus of medical management, the truth notwithstanding. Families, for their part, generally wanted the deceptive approach, and felt much more comfortable with it, and the deceptive approach, aided by the input and collusion of families, was seen as properly respecting their role and pivotal place in patients' lives, as well as their function as care providers in the community. Finally, allowing patients the option of denial in response to terminal or catastrophic illness was seen as the compassionate course, denial being seen as a natural, relief giving response to tragedy.

2.2 Bok's Criticisms of the Practice of Lying to, Deceiving or Not Being Candid with Patients

Half of Bok's criticism of the practice of lying to, deceiving or not being candid with patients involves identifying the benefits that are lost when such a practice is pursued, benefits that will be summarized in the next section below (IIC.) But she also spent considerable space challenging many of the factual claims of the paternalistic view. She had the advantage of referencing numerous studies which documented that patients do, in fact, want to hear the truth, (Robinson, 1976; Strull, 1985) as well as studies that showed patients were capable of understanding medical information, if such information was presented in lay persons' terms, boiled down to its essentials. (Bengler, 1974) She also accurately pointed out that few suicides occur as a result of truth telling, and the "death mechanism" argument was seen as lacking any empirical support.

The harms of the practice of lying to, deceiving or not being candid with patients were further held to be numerous and substantial, including the fact that deception tends to torture patients (Tolstoy's novella *The Death of Ivan Ilych* was often used to portray this), and leads to corrosive worry, especially in patients who, however reassured (falsely), do not get better and tend to get worse as their disease progresses. A patient's unanswered questions (asked or not) similarly led to such corrosive worry and undermined the hope that the deceptive approach otherwise sought to produce.

Bok also proceeded to identify other harms of the deceptive approach, including: (1) it becomes very difficult to proceed with and sustain potentially effective, aggressive treatment in a patient who is not aware of any good reason for it; (2) staff and family may well have different values and experiences than the patient, and may thus make choices for the patients that patients would not make themselves; and (3) patients, lacking accurate information, will be unable to do last things", e.g. making or revising their wills, saying their goodbyes, or taking a trip to Las Vegas to gamble away the inheritance of dishonorable children.

Beyond such harms, Bok relentlessly argued that the practice of deception often just did not work. The deceived patient eventually comes to suspect that he or she is being deceived, as aggressive treatment is attempted, and especially when the disease worsens. Eventually the patient realizes that he or she has been deceived, resents this, and loses all trust in the physician. As a further harm of the practice of deception, it was pointed out that family members who participate in such deception are quite likely to remember it when they themselves become ill and are reassured (appropriately or not). Thus the loss of trust in physicians becomes geometrical as family members come in for care. The charade not only does not work but, long term, will tend to undermine those values, e.g. trust in physicians, that effective medical practice depends upon.

2.3 Bok's Arguments in Favor of the Rule of Truth Telling

As one might expect from a Western thinker, Bok offered various points that keyed to the affront to autonomy and freedom that the practice of deception involved. In societies where such values are not strongly held, however, such arguments will not be telling. But Bok concurrently argued that the practice of deception involved giving up the opportunity to pursue various positive benefits that truth telling may further.

In part these benefits may be seen as a consequentialist rendition of the benefits of autonomy itself, and thus do not just involve appeals to some abstract principle of autonomy, but turn on the benefits that autonomy might have for any person. Here the fact that physicians and family members may have different values and experiences gets parlayed with the clinical observation that different people, if allowed to, make different choices about medical scenarios. Common examples of this are to be found in oncology where similarly situated patients, e.g. patients with metastatic disease, decide differently, some deciding to go to the local cancer research center for experimental responses to their disease, some electing to go to hospice instead, IF they are given the opportunity. These differences may also be captured in advance directives, if patients are given the chance, where their differing wishes are captured beforehand and allowed to dictate medical care when the time comes. Bok also offered various reflections about the personal value of autonomy that might best be rendered, in this context, as respecting the "superior man" that Confucian philosophy enshrines, a possibility that the deceptive practice violates. All such opportunities are, of course, lost if the patient is not given an accurate sense of his or her situation and prospects. (Wear, 1993)

Other potential benefits of truth telling were also emphasized by Bok, including: (1) truth telling enhances patient compliance and cooperation; how in fact do we think we can get patients to comply with often burdensome treatment if they are not aware of the reason for it?; (2) truth telling will equally tend to be pivotal in getting patients to accept aggressive diagnostic and therapeutic interventions, as well as provide a basis for a patient avoiding unnecessary or needless procedures; thus the possibilities for gaining acceptance of potentially effective aggressive treatment, as well as avoiding needless pain and suffering, and "bad deaths", will be enhanced; (3) studies have shown that anticipated pain is tolerated better by patients, and informed patients recover from surgery faster; (4) truth telling tends to counter the vulnerable and passive sick role that patients tend to lapse into, a state which may run directly counter to the mindset needed for extensive rehabilitation efforts where it is the patient's efforts that make all the difference; (5) misconceptions, false hopes and fears, needless pessimism and optimism, are best responded to by their antidote, the truth; and (6) by approaching the patient as a knowledgeable participant in care and decision making, the quality of everything from history taking, to self-monitoring and reporting, is likely to be enhanced. If left in ignorance, how can a patient even know what to watch out for, or report?

III. CONCLUDING REMARKS

A review of the literature on Chinese bioethics, especially that written by Chinese scholars and practitioners, shows that the practice of truth telling to patients, and of making them primary decision makers, generally meets with no enthusiasm, and is often impeached as running completely counter to the traditional placement of Chinese persons within their families, such families being seen as the appropriate decision makers. This surely has the benefit of recognizing the interrelatedness of people and the fact that, as one old saying puts it, "a man alone is in bad company". Further, the traditional Chinese approach is, it seems fair to say, vastly superior to the common Western tendency of approaching patients individually, without even requesting the presence of family members, as if the families' pivotal contribution to and support of care in the community is not absolutely necessary for much of the success of medical care. In this, it seems clear that the West has much to learn from the East, and the Eastern patient has many crucial advantages that the West seems to be ignorant of. (Brody, 1989; Kuczewski, 1996)

More starkly: as post-industrial capitalistic society emerges in full force in the east, one must wonder what other force might oppose and mitigate its destructive effects than the family. Such a society can and often is profoundly destructive of traditional societies and values, especially families. Equally, it can be absolutely overwhelming to the individuals caught in the belly of such a beast.

More prosaically, the first conclusory point is that however impressed one is by the consequentialist arguments of Bok et al., the last thing one should do is proceed in a fashion that deprives the Chinese patient of the support and succor that his or her family might provide. That support and succor can go way beyond anything that happens in health care and to the extent we appreciate the devastations that post industrial capitalist society can visit on people, anything that might undermine the traditional Chinese family should be opposed unwaveringly.

But there is clearly a tension here that is not going to go away, and will only increase with time. As the Chinese family becomes more geographically dispersed via the attempt by some of its members to participate in post industrial society, especially as its younger members leave for the cities by themselves and get more educated and worldly as a result, its authority and effectiveness are surely going to be challenged, and accommodations made if it is to remain viable. And for the sake of this paper, we should now pause to reflect on how Chinese medicine might react to and accommodate the challenges of modern medicine, some of which are captured in the consequentialist review of truth telling just offered. In this regard, a number of largely unargued suggestions as to how Chinese medicine might proceed regarding the practice of truth telling and its opposite will be offered by way of conclusion.

(1) At one extreme, it seems clear that the patient who clearly indicates that he or she wants the traditional Chinese approach in this area should be accommodated, viz. physicians should initially approach the patient's families with information and the families get to decide what should be done, as well as what is said. Whether this proviso should always be honored is not clear however. In the case initially provided, for example, even if the young lady indicated she wanted the traditional approach, one might be hard pressed, all things considered, not to want some clinician to advise this patient that some very important issues are involved in her care that the clinicians believes she should be appraised of, however much this should be done within the forum of the patient's family. For those who do not see the reasons for this cautionary note, I can only suggest they re-read my rendition of Bok's argument.

(2) At the other extreme, it surely seems that we must be realistic and flexible regarding the situation of increasing numbers of people/patients in contemporary Chinese society. That is: concurrently with the repeated caveat that Chinese bioethics must key to traditional Chinese society, its institutions and mores, various speakers at this conference have repeatedly noted that this society is becoming increasingly fractured and stressed in many ways. Aside from the generic effects of post industrial society, one can certainly read many of Mao's "programs", especially the Red Guards movement, as aimed directly at under-mining the Chinese family, and these have certainly not been without effect. One must thus suspect that, as in the West, many Chinese people/patients will come to health care essentially estranged from their families and in that situation reliance on traditional approaches seems simply inappropriate. Unfortunately, such individuals might need to be treated as in the West, as the atomistic, autonomous beings that they have become. (Appelbaum, 1987; Szczygiel, 1994) Some reaching out to re-constitute what they have lost might still be appropriate, but it would make no sense to proceed traditionally as if what no longer exists still does.

(3) It would seem safe to assume that most Chinese patients will fall into neither of the above categories, i.e. that most will neither indicate, upfront and unasked, that they want the traditional Chinese approach, nor be obviously estranged from their families. How should Chinese medicine approach these assumedly much more numerous patients? First, the basic suggestion would be that patients be queried, at the point of admission (or when taken on as a potential patient by a clinician) as to which sort of approach they would prefer. Arguably three choices will suffice: (1) the traditional approach, accurately described as involving informing the family without the patient present and allowing them to decide

both what will be done or not done, as well as what will, or will not, be said
to the patient, making clear that this would include situations where terminal
illness is present; (2) the Western approach where the patient is approached first
and gets to decide what is done or not, and what, if anything, his or her family
will be told; and (3) a joining of the two where the patient and the family are
approached and informed together.

By way of a further suggestion, it should be anticipated that many patients may
well have no clear preference among these possibilities, perhaps to the point of
becoming confused that such questions are even being asked and wondering what is
behind them. As often happens in the West, they may look instead to the physician
for guidance in making such a choice. How should the clinician respond? What,
from a clinical point of view, is the optimal situation, especially when the patient
has no preference?

For the reasons retailed in this paper, the third, mixed approach seems markedly
preferable over the traditional and Western approaches and arguably should be
routinely encouraged by clinicians to the extent they are given the opportunity to
do so. Bok's arguments are offered as quite robust regarding the problems and
potential harms of the unleavened traditional approach, and a further presumption
is that this will become increasingly apparent over time within Chinese health care.
Increasing numbers of people will see what is going on and it will not work. And
many important benefits will otherwise be lost whether it works or not. Similarly,
though one would need to give Chinese patients the option of the autonomous,
confidentiality protecting Western approach, it seems appropriate, given all that has
been said, for clinicians to encourage upfront, substantial involvement of patient's
families in the informing and decision making processes. This not only provides
for the chance of such families providing the support and succor that may well be
crucial for the patient, but avoids alienating them at a time when their full, active,
knowledgeable engagement may well be essential to the care of the patient.

A shared model of informed consent and decision making is thus being offered
here, and a basic element of the Chinese tradition might be used to articulate
it. That is: however much the family is encouraged to be and stay involved, to
contribute to decision making in a knowledgeable way, and so forth, the patient
should be elevated to the status of "key person" as far as medical information and
decision making go. Too many harms accrue, too many benefits are lost, within
the contrary practice of deception; that at least, as far as this paper goes, seems to
be the dominant calculation. Patients would certainly have the option of deferring
to the judgments of their family members, and if many Chinese writers are right
about the primacy of the family in Chinese life, then patients will surely do so. In
all this, nothing need change except that the patient would be offered the truth and
be allowed to fashion his or her own response to it, however conditioned by family
input. Equally, within such an approach, the sensitive clinician will have ample
opportunity to recognize the patient who really prefers to be reassured rather than
told the truth, as well as the patient who suspects the truth, but does not want his or

her nose rubbed in it. Our experience in the West is that patients tend to hear what they want to and are quite capable of filtering out unwanted and unpalatable news.

Will such a change to medical practice tend to undermine the core place of the family in Chinese society? If that place is as robust and primary as many of our Chinese colleagues think it is, it is not clear how or why this would occur. The traditional mechanisms of deference to the family can still occur, the only difference would be that the patient would be knowledgeably deferent, and might presume to make his or her own perhaps different case within the counsels of the family. It seems hard to believe that this has not always occurred in other areas, from selecting a spouse, to choosing a livelihood.

And a final thought: as China itself enters the post-industrial age, with increasing numbers of its citizens, especially younger ones, leaving the immediate confines of their families to seek their fortunes, and getting much better educated so as to be able to compete in their chosen professions, some accommodation to this by the traditional Chinese view would seem essential...nay unavoidable. Add to this the ubiquitous presence of the internet and all the information it provides, and successfully deceiving patients about "shadows" on their lungs, or whatever, becomes much less feasible. It increasingly just did not work in the West, and this seems to be as likely in the East. So some modification of the traditional Chinese view would seem to be absolutely unavoidable. The compromise solution proposed here is that the traditional Chinese practice of lying to or deceiving patients be abandoned in favor of a goal of truth telling to patients, but the pursuit of this goal should be placed firmly, if possible and allowed, within the bosom of the Chinese family, with the patient being elevated, if he or she is able and willing, to the status of "key person" regarding those decisions that relate directly to the nature of his or her own medical care. To do otherwise may well, as Bok's arguments suggest, simply harm patients in numerous ways, fail to pursue important goods, and confine Chinese medical practice to a hide bound approach that lacks the sensitivity and flexibility required.

REFERENCES

Annas, G. (1975). *The rights of hospital patients*, Avon Books, New York.

Appelbaum, P.S., Berg, J.W., Lidz, C.W., & Parker, L.S. (1987). *Informed consent: Legal theory and clinical practice*, Oxford University Press, New York.

Bengler, J., et al. (1980). 'Informed consent: How much does the patient understand?,' *Clinical Pharmacology and Therapeutics*, 27: 435–40.

Blackhall, L., et al. (1995). 'Ethnicity and Attitudes toward Patient Autonomy,' *Journal of the American Medical Association*, 274 (10), 820–825.

Bok, S. (1978). 'Lies to the Sick and Dying,' *Lying: Moral choice in public and private life* (pp. 232–255), Vintage Books, New York.

Brody, H. (1989). 'Transparency: Informed consent in primary care,' *Hastings Center Report*, 19: 5–9.

Chan, H. (2004). 'Informed Consent Hong Kong Style: An Instance of Moderate Familism,' *Journal of Medicine and Philosophy*, 29 (2), 195–206.

Cong, Y. (2004). 'Doctor-Family-Patient Relationship: The Chinese Paradigm of Informed Consent,' *Journal of Medicine and Philosophy*, 29 (2), 149–178.

Fan, R. and Benfu L. (2004). 'Truth Telling in Medicine: The Confucian View,' *Journal of Medicine and Philosophy*, 29 (2), 179–193.

Fan, R. & Tao, J. (2004). 'Consent to Medical Treatment: The Complex Interplay of Patients, Families, and Physicians,' *Journal of Medicine and Philosophy*, 29 (2), 139–148.

Fielding, R. & Hung, J. (1996). 'Preferences for Information and Involvement in Decisions During Cancer Care Among a Hong Kong Chinese Population,' *Psycho-Oncology*, 5, 321–329.

Katz, J. (1984). *The silent world of the doctor and patient*, Free Press, New York.

Kuczewski, M.G. (1996). 'Reconceiving the family: The process of consent in medical decision making,' *Hastings Center Report*, 26, 30–37.

Meisel, A. (1979). 'The "exceptions" to the informed consent doctrine: Striking a balance between competing values in medical decision-making,' *Wisconsin Law Review*, 2, 413–88.

Parsons, T. (1975). 'The sick role and the role of the physician considered,' *Milbank Memorial Fund Quarterly*, 53, 257–277.

Ramsey, P. (1970). *The patient as person*, Yale University Press, New Haven, Connecticut.

Ravitch, M.M. (1978). 'The myth of informed consent,' *Surgical Rounds*, 1, 7–8.

Robinson, G. & Merav, A. (1976). 'Informed consent: Recall by patients Tested Postoperatively,' *Annals of Thoracic Surgery*, 22, 209–212.

Strull, W.M., Charles, G., & Lo, B. (1985). 'Do patients want to participate in medical decision making?' *Journal of the American Medical Association*, 252, 2990–2994.

Szczygiel, A. (1994). 'Beyond informed consent,' *Ohio Northern University Law Review*, 21, 171–262.

Wear, S. (1993). *Informed Consent: Patient Autonomy and Clinician Beneficence within Health Care*, Kluwer Academic Publishers, Dordrecht, Holland.

Wear, S. (2004). 'Informed Consent,' in G. Khushf (Ed.), *Handbook of Bioethics: Taking Stock of the Field From a Philosophical Perspective* (pp. 251–290), Kluwer Academic Publishers, Dordrecht, Holland.

CHAPTER 7

ON RELATIONAL AUTONOMY
*From Feminist Critique to Confucian Model
for Clinical Practice**

SHUI CHUEN LEE

National Central University, Taiwan

The Twentieth Century is often characterized as a time when autonomy triumphed in biomedical practice. However, its central position has been challenged in the last decade from clinical and philosophical quarters. While empirical findings reveal that most patients, especially those facing serious medical decisions at the bedside, are not making any real autonomous decisions and need and want help and guidance from others. Feminists argue that the traditional individualistic concept of autonomy is a kind of abstraction that deprives and betrays the true identity and needs of the patient. They propose the concept of relational autonomy to restore the patient to her embedded situation as a member of her community. This concept gives due regard to a person's personal identity and brings in the help of related persons, family members, friends, etc. While there are a number of different versions of the concept of relational autonomy, these treatments are not very satisfactory. This paper argues that the Confucian concept of a person can provide an answer to most of the intricate problems with this concept. A Confucian analysis of the nature of our moral experience supports the primacy of autonomy for a moral being and reveals the relational character of such a being. In doing so, the concept of relational autonomy is aligned with the basic idea of autonomy in the Kantian sense, and the concept of person is being reinterpreted according to the Confucian concept of a person. According to this version of the concept of autonomy, a person's family is her main source of personal identity.

---

* This essay is one of the papers produced by my project on "The Personhood and Moral Status of Embryos: The Challenge of Human Genome," which is a three-year project (2002–2005) supported by the National Science Council, Taiwan. Its earlier versions have been presented in Beijing International Bioethics Conference, Beijing, January 2004 and in the Seventh World Congress of Bioethics, Sydney, November, 2004.

S.C. Lee (ed.), The Family, Medical Decision-Making, and Biotechnology, 83–93.
© 2007 *Springer.*

Such a concept emphasizes the family relationship and the concept of autonomy is renamed in the process; it can now be called "ethical relational autonomy." We can then extrapolate this concept into a Confucian model for clinical consideration and application.

In the clinical setting, the promotion of informed consent and efforts to respect the autonomy of the patient has been very fruitful in moving the practice of medicine towards patient-centered development. However, many studies also show that the practice of autonomy often leaves the patient in a helpless situation when her family and other acquaintances are excluded, and as a result the patient often falls prey to the implicit manipulation of medical professionals. In Chinese societies like Taiwan, which follow the Chinese tradition, the family plays an important role in medical decision making. Not only does the patient often ask and need the advice of family members, but physicians too, ask for the family's opinions and decisions for treatment options. This is expressed in the saying that when a family member is sick it is as though the whole family is sick. Every family member including the patient feels that making medical decisions is not only the patient's business but part of the family's business. Elderly patients, more often than not, will take the interests of the family as a whole into account when considering treatment options and tend to make choices that result in fewer burdens for their families, often at the detriment of their own health.

Over the last decade or so, feminists have contributed greatly to the critique of the traditional concept of autonomy and have broken new grounds for a reconfiguration of this important concept. In what follows, I shall first provide the feminists' critique of the traditional concept of autonomy, and their suggestions for the concept of relational autonomy. I shall then strengthen the concept of relational autonomy both by grounding its root in morality and justifying how the identity of a person relates to her social relations and, in particular, her familial relations. Finally, I try to refine the notion of relational autonomy into a Confucian notion of "ethical relational autonomy" and show how it can be deployed at bedside consultations.

I. FEMINIST CRITIQUE OF THE CONCEPT OF INDIVIDUAL AUTONOMY

Traditional individualistic concepts of autonomy stress the independence of the subject and are often in opposition to all types of human relationship. This has led them to be contested vigorously by feminists who tend to regard relations as essential for moral judgment and actions. Traditional individualistic concepts of autonomy as they have been presented in the main stream of bioethics have been subjected to five kinds of critique, namely, symbolic, metaphysical, care, postmodernist and diversity critiques (Mackenzie & Stoljar, 2000, pp. 3–31). These critiques assess the notion that the moral agent is as an asocial, atomistic and abstract individual, who can be completely independent of all social ties. However, in spite of its defects, at least some feminists do not think that we ought to throw out traditional concepts of autonomy completely, but only that we need to reconfigure them.

Traditional concepts of autonomy are derived from two types of theories of autonomy. The most popular type is called a procedural theory of autonomy and the other type is called a substantive theory of autonomy. Marilyn Friedman characterizes procedural theories as follows:

> According to a procedural account, personal autonomy is realized by the right sort of reflective self-understanding or internal coherence along with an absence of undue coercion or manipulation by others (Friedman, p. 40).

And she points out that such a view of autonomy does not imply living substantively in any particular way. Mackenzie and Stoljar concur and summarize procedural theories as follows:

> On procedural, or content-neutral, accounts, the *content* of a person's desires, values, beliefs, and emotional attitudes is irrelevant to the issue of whether the person is autonomous with respect to those aspects of her motivational structure and the actions that flow from them. What matters for autonomy is whether the agent has subjected her motivations and actions to the appropriate kind of critical reflection (Mackenzie & Stoljar, 2000).

However, feminists are keen to point out that procedural theories, which take autonomy to be an act of critical reflection, cannot distinguish certain subtly conceivable forms of non-autonomous actions where agents seem to act autonomously but are subjected substantively without self-awareness. Such theories are criticized as at best only setting out some of the necessary conditions of being autonomous because how the critical reflection is carried out and whether the ground of critique is sufficiently enlightened affects the result of critical reflection, especially whether the agent is autonomous.

Substantive theories of autonomy are of two kinds, namely weak and strong substantive theories. Mackenzie and Stoljar summarize these as follows:

> The former rejects the content neutrality of procedural theories by requiring specific contents of the autonomous preferences of agents. The latter reject content neutrality by suggesting further necessary conditions on autonomy that operate as constraints on the contents of the desires or preferences capable of being held by autonomous agents (Mackenzie & Stoljar, 2000).

The weak versions usually place the constraint of the agent in self-worthiness, self-respect and self-trust, and try to explain how the real situation can deprive the agent's autonomy through unfavorable socialization. However, these weak versions do not state the substantive content – say, what worthiness is, more specifically. The strong versions require that the agent be competent or has the capacity to differentiate between right and wrong – that is to say, an agent must be a rational self-legislative actor. They are thus regarded as a new version of the Kantian account of autonomy (Wolf, 1987). Such theories make the notion of autonomy go beyond a purely procedural one and, in fact, are reconfiguring it towards a relational concept of autonomy.

II. FROM INDIVIDUAL TO RELATIONAL AUTONOMY

From a feminist point of view, the traditional concept of autonomy may be too masculine and traditional societies offer few chances for women to be independent. However, the traditional notion has a definite moral implication for the liberation of women from unjust social settings. Though it neglects the real issue, namely, the personal relations inherent in real life situations, the language of autonomy nevertheless provides conceptual tools for combating anti-gender and anti-racial discrimination. Hence, the major trend seems to favor a reconfiguring of the concept rather than throwing the baby out with the bath water.

One of the strongest arguments for the expansion of traditional concepts of autonomy towards a relational concept of autonomy is that the identity of a person cannot be separated from her personal attachment with various social sectors. One is a member of a certain family, of a certain ethnic group, belongs to a certain social sector through education, occupation, association, and so on. In other words, a person is not living in a void, an asocial or atomic existence. Her identity must come with all such relationships. Without knowing one's identity, one cannot achieve self-knowledge. Without understanding one's identity, one cannot really be self-conscious, and true to one's own self, and thus one cannot be autonomous. Hence, autonomy must be relational.[1]

This of course leads to the further question of what kinds of relations are going to be included in the concept of relational autonomy. On this subject, feminists seem quite divergent. Most seem to be purposely vague about the kind of social relationship that ought to be included in this concept. Some mention friendship while some talk about kinship. Maybe it is due to the fact that this concept is still being investigated and there is no consensus yet. It is useful to point out that one's authentic self and thus autonomy must be considered within the social context one is born into. However, as most feminists realize, we cannot identify relational autonomy with social connectedness without qualifications. Social settings, whether favorable or not, can both promote and frustrate our autonomous performance or achievement.[2] It is, as I see it, a certain kind of unnecessary tattoo for feminists using a kind of non-hierarchical model like that of friendship to present the case for relational autonomy, while avoiding the family model. For many feminists, the family is the base and often the shielded private sphere where oppressive social institutions upon women are being propagated and routinely carried out. However, they seem to forget that one's identity is often related to the close and intimate relations of parents and children, sisters and brothers. Further, this familial relationship is not only a social relation that one is born with, but it is also deeply ingrained within our conscious and sub-conscious levels. This relationship can be counted as one among other social relations but it is definitely the most influential of our social relations – it even influences the worst side of our person-ality. We have no reason to side step it and ignore its existence. It seems that in doing so, we are somehow making a "bad faith" type of mistake in regard to our authentic self.

I shall try to clarify the role the family plays in our personal identity. Before that, I would like to dive more deeply into the roots of morality in order to better ground this account of relational autonomy and show how our personal identity figures in our autonomous acts.

III. PHILOSOPHICAL EXPLICATION OF AUTONOMY: FROM KANT TO CONFUCIANISM

The modern significance and prominence of autonomy no doubt stems from Kant's moral thinking and, hence, it is worthwhile to return to his conception of it in order to drive the wedge back into the root of morality. However, it has been shown that the present use or uses of autonomy have been largely deflected from Kant's strict use and are, in fact, more often than not against his formulation (Secker, 1999). For Kant, autonomy comes from our free will and is in contrast to what he called mechanical or natural causality. However, the causality that comes from free will does not mean that it is arbitrary. Kant emphasizes that it is by all means law abiding, though the law comes from free will itself. Hence a moral agent is a self-legislating person. For Kant, a moral agent is not anyone who acts from her arbitrary will or choice. One's moral act is dictated by the categorical imperative, which in some sense is not so much a personal or private choice but a decision as a rational being or from the stand point of all rational beings. Though Kant's formulation is usually regarded as purely formal, that is, without any substantive delineations of the content of a moral act, the categorical imperative contains substantial implications to serve as guidance for our moral judgments.[3] Or, it is charged as smuggling the substantive valuing of freedom as supreme (Engelhardt, 1996). Inevitably, Kant must mean by morality that we have certain moral concerns with our acts. This moral concern is nothing other than the concern of our wellbeing and the wellbeing of other rational beings; otherwise we cannot understand how and why a law-abiding act is moral. Kant emphasizes that our actions must be unconditional or we must act from duty alone if our actions are going to have moral value. But, this does not make our actions merely formal; it only indicates that a moral act must be an act solely for the sake of duty or flowing from the unconditional command of our moral law. In fact, Kant did try to employ his categorical imperatives for analysis of certain moral issues, such as lying, suicide, helping others and self cultivation, and he also indicated that certain acts were immoral according to our moral law. It means that the categorical imperatives are by no means merely formal, and they contain certain substantive contents that make them moral laws. It is clear that Kant has certain moral concerns with his moral laws although we may argue about whether his analysis is satisfactory. In a nutshell, Kant requires that a moral act be an act from a moral agent who is self-legislating and autonomous. Unfortunately, Kant stopped short of giving a description of the type of person he has in mind when framing a person as one that has intrinsic worth because of her endowment of a free will. To make this concept more explicit, let us turn to Confucianism.

Confucianism fully accepts the Kantian concept that a moral act must be some act circumscribed by the moral mind or free will or practical reason in Kant's sense. For Confucianism, the moral mind or moral consciousness is the origin of our moral endeavors. Confucianism starts with the mind that cannot bear the suffering of others (*Mencius* 2A: 6).[4] It is the natural endowment we are born with and which usually reveals itself when we encounter injuries occurring to other living beings, especially serious injuries to a human being, for example, a detrimental injury to a child. When this moral alarm or moral consciousness appears, it moves us into the realm of morality. It has two characteristics. First, it points to a definite direction, the direction of a demand or an urge that such calamities should not happen. It commands us to act to relieve the calamities. Thus, it carves out the realm of morality, where we have to act either morally, that is following the self imposed request of our moral consciousness, or act immorally by neglecting this internal command and going astray. Confucianism sees this spontaneous act of our moral mind as constitutive of our moral realm. This internal consciousness is thus primordially legislative and self-legislating as it gives itself a definite direction. Being unconditional and hence universal, it is in Kant's sense a self-legislation in the form of a moral law. According to Confucianism, our moral principles, such as the principles of *ren* (benevolence), *yi* (justice), *li* (rituals), and *chi* (moral awareness), evolve from this primordial moral consciousness. This moral consciousness embodies our autonomy. It signifies our internal humanity and is what makes us an end in itself. Human beings, or for that matter, in Kant's terminology, rational beings, are special because they have moral experiences and moral concerns about their own actions. Usually, we do not assume that animals have concerns about moral matters, as we cannot be sure that they have the kind of moral capacities as we do. Our moral concern reveals that morality comes from our special constitution, which we typically call free will, or practical reason, moral mind or human nature (it does not make much difference here). Confucianism claims that this moral mind is what makes a human being have moral worth. It makes us a moral agent and thus a being of utmost value. Its command is an imperative that demands us to equally treat others as ends in themselves.

IV. MORAL RELATEDNESS AND FAMILY RELATIONSHIP

The second characteristic of the primordial moral act is that our moral consciousness is at the same time orientated to others. It is a moral consciousness that expresses unconditionally our concern for the suffering, or well-being or happiness of others.[5] Acting autonomously is never primarily a concern with our own personal happiness, nor personal interest, and never involves ignoring the suffering of others. It would be against our experience and concept of morality if autonomy were solely a concept concerning the agent herself. Thus, autonomy as it reveals itself through our concern for the suffering of others is by its nature a notion of relatedness, a moral kind of connectedness. It is precisely Kant's view that morality is not personal or species specific. He tries to make this point when he eschews all our subjective personal preferences, nay, all human beings' special preferences from our moral acts. However, this does not mean that the autonomous person is an atomic individual at all; our autonomy for Kant expresses our concern for all moral agents.

From the Confucian perspective, a moral act is primarily concerned with the suffering of others. Sometimes, though rarely, it can also direct moral concern towards our own suffering. Hence, the notion of harm naturally comes to the forefront in our moral endeavors. It is no wonder that both Confucius and Mencius took concerns of the happiness of others (benevolence) and sufferings as the primary targets of moral discussion as they are in fact two sides of the same coin.[6] Benevolence and harm are invariably other-orientated. Thus, moral agents are first of all morally connected. As moral agents, they form a moral community, which is bound by moral laws, not social norms or contracted rules. It seems evident that we are morally bound to all human beings without any prior contracts of any kind. We naturally feels our duty, especially our duty of humanity, towards every human being on Earth no matter whether we are acquainted with them or not. For Confucians, it extends even further, for we cannot withhold our concerns for other living things and limit our concerns to our own species. Every entity that can feel and be harmed is within our circle of empathy.

Because human beings are finite, our moral practice invariably starts from what is most closely at hand although our moral concern can be unlimited. Everyone is born to a mother or born into a family in the strict sense. The overwhelming majority of us are raised within a family. If we take into consideration single parents, adoption and legal guardians, we can say that everyone is raised within a family. It seems obvious that without a family, except for some very special cases like the "wolf child," no infant can survive more than a few days. One of the merits of feminist philosophy is to point out that some of our most intimate relations are part and parcel of our personal identity and self-identity. According to Confucianism, our family relationship is the most intimate and eternal one of all our relations. In a sense, the experience of our earliest years is forever imprinted with us and we will never be completely detached from it, however hard we may try to get rid of it. It does not mean that we cannot change our family identity, say, set up our own family or deny our former affiliations. However, such moves like setting up our own family usually result in an identity that consists of a mixture of the two families.

Since the locus of the family is so closely connected with our lives, this initial and closely bounded association is where our moral acts start. The binding force of the family is more than a general moral community, but also consists of the reciprocal relationship between family members. For as intimate members living under one roof over a certain period of time, each is in some sense not only emotionally connected to the others, but is also bound by reciprocal responsibilities, which cannot be shared by outsiders. In a sense, family members are amalgamated into one, as an organic whole. In Confucianism, the relationship between family members is morally defined in terms of mutual and reciprocal responsibilities rather than a hierarchy involving domination and subjection. For example, parents and children are bound by benevolence and filial piety, husbands and wives are bound by sharing family duties (differentiation or division of labor), brothers and sisters are bound by love and respect.[7] Thus, the close and intimate family relation bears significantly upon our duties and responsibilities to each other within the family. We must first take into the needs and happiness of our children or parents. This kind of partiality is, in fact, a version of universal and impartial justice. Confucians regard everyone as having such a special and important duty to one's family members.

Hence, one is not only a moral agent but is also related to others within her family. This notion of autonomy, which views one as being morally related to other family members, suggests that we have the concept of autonomy embedded in our family relationships. Because of its ethical component, this concept of autonomy can be called ethical relational autonomy. We may now explore further the formation and implications of such a Confucian concept, which are central to our bioethical deliberations.

V. A CONFUCIAN MODEL OF ETHICAL RELATIONAL AUTONOMY

Generally speaking, a human being is born with two kinds of endowment. One is her natural endowment, which includes her rational and moral characteristics. Another is the human-made environment, that is, human culture, which signifies her existential situation. The former provides us with moral consciousness, which is as we said above, constitutive of our moral realm. The other endowment, though artificially made by the human being herself, is nevertheless part and parcel of our lives. We are born into a society, a society with various kinds of values and relationships. The kind of moral capacity that we are born with operates within these human environments and what happens after we are born into these environments constitutes our personality and our personal identity, together with our moral relationship with others.

We can describe the human environment broadly as our social environment and the constituents of social relations. Some of the formal structures may be called social, economic, and political status or relations. Some may be referred to as ethnic, racial, gender, family relations. Of these, the closer the relation, the more it bears upon our personal identity. The day-to-day experience is what our identity is all about. Thus, the natural familial experience is immensely constitutive of our personality. In fact, many of those broader social relations and their influences come to the subject through family encounters. Accordingly, we regard the family as inalienable from one's personal identity and wellbeing. Conversely, the wellbeing of other family members or our family as a whole cannot be completely indifferent to our well-being and identity. Family is our somewhat non-voluntary center of intimate association with others. It is different from our voluntary associations with friends or working partners. The latter are usually much less intimate or persistent, and often frivolous. Marriage is another source of great influence on our personal identity. A married couple live at least for a certain period of time in very close and intimate relationship and if they have children then it binds them more closely in the usual kind of family ties. These two types of strong and intimate relations, namely marriage and the parent-child relationship, constitute much of our personal life history and personal identity. Their effects invariably build into our character or personality. We may, of course, subscribe to a broader notion of familial relations and take any type of intimate relationship into our consideration. Thus, the consultation of family in cases of patient treatment may involve persons other than the legal family members. Such important and inalienable relations are called ethical relations in Confucian terminology.

Though our moral commitments are family specific, our moral concern is basically non-specific and is based upon our universal and unconditional concern of the suffering of others, it is rational and self-critical. Hence, ethical relational autonomy is not only a manifestation of our moral consciousness and what makes us a person of moral worthiness, it is also a spring or ground where we can put our value or judgment under critical scrutiny. The ultimate ground for critical evaluation of present institutions or norms of our society or community lies with our concern of the harm or well-being of others.

According to Confucianism, all living things are members of a big family. Ontologically speaking, we cannot be separated from our family, our society, our nationality and the whole human race. Furthermore, we cannot give a full and satisfactory account of our personal identity without taking into the fact that we are also part of the whole chain of living things. If we take into consideration the advice of land ethics, we cannot separate ourselves from Mother Earth as a whole. As such, they are objects of our moral concern and necessarily figure in our relationally autonomous actions. What are we if not some moral apes that happen to thrive on Earth?

VI. SOME PRACTICAL IMPLICATIONS OF ETHICAL RELATIONAL AUTONOMY: THE ROLE OF FAMILY IN BIOETHICAL AND MEDICAL DECISIONS

Our concept of ethical relational autonomy builds upon relational concept of autonomy espoused by feminists and has the blessing of the latter's critique of individual autonomy. It also has two significant advantages over the latter. First, it is secured by going to the root of morality, and thus offers a moral justification for its being the true meaning of what our moral concerns designate. The basic elements of the procedural theories, such as critical reflection, competence, self-consciousness, are part and parcel of being autonomous. It also gives full attention to the special aspects of relatedness of human beings. It is in a sense specifically substantive and thus has the further advantage of being more workable with the real situations that one encounters in life. In the remaining space, I shall try to show how it provides a model for bedside consultation.[8]

For bioethical problems in a Chinese society, the family is the background of our personal acts and decisions. Many of the issues, in fact, concern the family as a whole, such as genetic information, privacy and confidentiality. Even in cases of choosing treatment options, the opinion and participation of the family is essential. The family as a unit is the object of the application of the principle of respect of autonomy. The agent is acting on behalf of his family. When one is alone facing others, one seems to act individually, but one is, in fact, acting as a representative on behalf of her family. Harms and benefits, or interests and preferences, are bestowed to family and family members as a whole. Individual autonomy is but a heuristic way of speaking.

In the clinical situation, if we are not satisfied with the individualistic concept of autonomy, we face the difficulty of which related person or persons should participate in cases concerning an individual patient's choice in medical affairs, and what can we do if there are conflicts between the parties, especially when the patient is at odd with family members. Relational autonomy implies that intimate persons should be consulted, especially in cases when the subject is incompetent. However, it is the usual practice in the west that the patient herself has the final say, and often, even in cases of an incompetent patient, the related party's decision has to take into consideration the best interests of the patient. Now, if we accept an ethical, relational concept of autonomy, it means that autonomous decision making is a family affair or a collective right. When one is sick, it is regarded as a sickness for the whole family, and usually the best interest of the patient is also the best interest of the family. In other words, the patient and the family members have a close and common interest together, and thus have duty and claim to each other. In this vein, the consultation of the family and collective decision is final. There is usually no pain of deception or concealment. The patient is protected and supported by family members and blessed in unfortunate situations. When patient and family have conflicts, very often the patient would choose to do what is less burdensome to her family, and we suggest that the first resolution should be to seek dialogue and compromise. If this conflict cannot be resolved, the patient's last wishes are usually respected. This model relieves the burdens of physicians in medical decision making because the family is usually a competent unit and can make competent treatment decisions. The physicians can get the help of the family in understanding the real wishes of the patient and in communicating certain information, such as bad news or treatments and prognosis. However, the physicians and nurses, and very often the social workers too, somehow form part of an extended community that includes the patient and her family. They have the professional duty to see that the best interests of the patient, as well as her autonomous choice be protected. Very often, the family also needs help making critical decisions concerning the patient, and our model gives them proper and due concern.

NOTES

[1]   This is a common theme and argument of many feminists and their writings, cf., for example, Diana Tietjens Meyers (2002).

[2]   Meyers gives a very good examination of this theme in her paper with the notion of "intersectional identity" in her 'Intersectional Identity and the Authentic Self? (Meyers, 2002)'

[3]   Cf. how Kant explains his categorical imperatives implies that we have certain definite duties to ourselves and to others in his *Groundwork of the Metaphysic of Morals*.

[4]   Cf. *Mencius* 2A: 6. For Confucius, his key notion is *ren*, and in the *Analects*, there are similar statements stating that *ren* is our moral consciousness and concerns of others. Please refer to (Lee, S.C., *Confucian Bioethics*).

5 For the moment, we limit it to the concern of other human beings, though Confucian cares for the well-being of all living things. The implication of this point will be expressed latter.
6 That harm and benevolence are on a continuum is layed out in Tom L. Beauchamp and James F. Childress (2001), *Principles of Biomedical Ethics*. That our moral talk begins with concerns of others seems a universal phenomenon. Not only Confucians, the basic notion in the Hippocratic oath is "do no harm," whereas the notion of autonomy comes much later and it is through Kant's work that it became prominent. Part of the cause is the wide spread of liberal individualism both in the political and moral life of the modern society.
7 The use of male terminology in presenting the mutual responsibilities in Chinese may give some suspicion of patriarchal bias. However, sometimes it is more of an economic way of expression and there is no reason to say that the same kind of moral relationship does not hold between mother and son, or father and daughter, or brothers and sisters. The so-called "bonds of domination and subordination" between emperor-officials, father-son, and husband-wife simply never appears in Confucius and Mencius, nor in any pre-Chin Confucian writings. Both Confucius and Mencius, especially the latter, put forward the kind of moral reciprocal relationships between the so-called five ethical relationships of father-son, emperor-official, husband-wife, elder-young, friend-friend, as loves, justice, coordination, respect, trust respectively. Responsibility is reciprocal. It is never one-sided domination and subjection. Even the short and somewhat latter document of the "Book of Piety," documents Confucius as saying explicitly that sons and officials should counter their fathers or emperors if the latter are wrong. The bonding way of specifying family relation is commonly recognized among Chinese scholars as a distortion by the thoughts of Chinese legalism, which is basically a philosophy supporting authoritarian rules of the emperor. Unfortunately, it wields with political power that lends support to the feudal states and patriarchal society for the last twenty thousand years in China.
8 I have developed a model of bedside consultation with the concept of ethical relational autonomy in a collective project with my colleagues in medical schools and physicians in Taiwan in a book in press. We have visited many hospitals and discussed cases with those in charge and concerned. I have benefited greatly with their experience and questions. The following is an outline of the model and how it works.

REFERENCES

Beauchamp, T.L. & Childress, J.F. (2001). *Principle of Biomedical Ethics*, 5th Edition, Oxford Press, Oxford.

Engelhardt, H.T., Jr. (1996). *Foundations of Bioethics*, Oxford University Press, New York.

Friedman, M. (2002). 'Autonomy, Social Disruption, and Women,' *Rational Autonomy: Feminist Perspectives on Autonomy, Agency, and the Social Self* (pp. 151–180), Oxford University Press, Oxford.

Jecker, B. (1999). 'The Appearance of Kant's Deontology in Contemporary Kantianism: Concept of Patient Autonomy in Bioethics,' *Journal of Medicine and Philosophy*, 24.1, pp. 43–66, Swet & Zeitlinger, Dordrecht.

Kant, I. (1964). *Groundwork of Metaphysic of Morals*, H.J. Patron (trans.), Chapter 2, Harper & Row, New York.

Lee, S.C. (1999). 'Confucian Biothics,' Chapter 2, Legion Monthly Press, Taipei.

MacKenzie, C. & Stoljar, N. (Eds.) (2002). 'Autonomy Refigured,' *Relational Autonomy: Feminist Perspectives on Autonomy, Agency, and the Social Self* (pp. 3–31), Oxford University Press, Oxford.

Mencius (1970). *Mencius*, D.C. Lau (trans.), Penguin Books, Suffolk.

Meyers, D.T. (2002). 'Intersectional Identity and the Authentic Self ?,' *Relational Autonomy: Feminist Perspectives on Autonomy, Agency and the Social Self* (pp. 151–180), Oxford University Press, Oxford.

Wolf, S. (1987). 'Sanity and the Metaphysics of Responsibility,' in F. Schoeman (Ed.), *Responsibility, Character, and the Emotions*, Cambridge University Press, New York.

CHAPTER 8

REGULATING SEX SELECTION IN A PATRIARCHAL SOCIETY
Lessons from Taiwan[1]

WENMAY REI

National Yang Ming University, Taiwan

I. INTRODUCTION

Unless born with a physical abnormality, a baby is usually known first by its sex. To people shopping for newborns, consumer markets offer congratulation cards, clothes, and toys that differ depending on whether the baby is a boy or a girl. Societies have different customs to celebrate the birth of babies, but almost all these societies have one thing in common: each differentiates its customs on the basis of sex. Likewise, parents may have different ideals about raising a son or a daughter and may act upon them accordingly.

Against this backdrop, people often seek to increase their chances of bearing a son through arrangements that range from special diets, coital timing, and specific positions of intercourse to immunologic manipulation, but the reliability of these methods ranges from the dubious to the superstitious (Jones, 1992, pp. 4–6; Holmes, 1995, pp. 156–157).

Assisted reproductive technologies (ART) significantly elevate the chances that sex selection will be successful. The development of ultrasound screening, amniocentesis, chorionic villi sampling (CVS), and blood testing make it possible for people to learn the sex of their fetus before birth. Because these techniques identify the fetus' sex only after the fetus exists in the mother's body, the only certain way to avoid a child not of the "right" sex is to abort the fetuses of the "wrong" sex.

But sex selection does not necessarily involve abortion. In 1924, researchers discovered the existence of sex chromosomes, and, in 1971, it was learnt that males typically have a greater number of X-bearing sperm than Y-bearing sperm (Sumner, 1972, pp. 231–232). Ever since then, different sperm-sorting technologies such as gradient methods and flow cytometry have been developed, and, as a result, the likelihood of bearing a son has, according to some researchers, attained a success rate of approximately 90% (Björndahl & Barratt, 2002, pp. 2–9). Because these

95

S.C. Lee (ed.), The Family, Medical Decision-Making, and Biotechnology, 95–111.
© 2007 *Springer.*

techniques intervene by ensuring that the "right" sperm fertilizes the egg before any embryo is harmed, they avoid the criticism that is often leveled against sex selection associated with both abortion and the destruction of embryos. In recent years, researchers have used pre-implantation genetic diagnosis (PGD) to investigate the genetic makeup of a pre-embryo produced by in vitro fertilization (IVF) before the pre-embryo is transferred to a woman's uterus. In this context, PGD raises the accuracy of the selection to nearly 100% (Björndahl & Barratt, 2002, pp. 9–10).

Are parents allowed to choose their children's sex before they are born? In many countries, the answer is a resounding no. For instance, Article 14 of the Council of Europe Convention for the Protection of Human Rights and Dignity of the Human Being with Regard to the Application of Biology and Medicine specifically prohibits parents to use medical technology for the selection of their child's sex unless the purpose is to avoid a serious hereditary sex-related disease.[2] Thus, many European countries ban sex selection that is based on non-medical reasons.[3] Recently, the Human Fertilisation and Embryology Authority of the United Kingdom, after a year-long review and public consultation, also recommended the same position.[4]

When we compare all the choices that people are allowed to make about reproduction, we see that a ban on parents' selection of their child's sex is significant in two ways. First, the target of such regulations is people's motivation: instead of banning the technology of sex selection altogether, the government draws a line between sex selections that are based on medical reasons and those that are not. Second, by drawing the line according to the presence or the non-presence of medical reasons, governments rely on the medical profession for the justificatory power to decide whether sex selection is permissible.

However, in a society where sons are predominately preferred, and where cultural and legal institutions still favor males, a regulatory system that aims to police sex selection is both tainted and constrained by the sexism embedded in the society's formal and informal social structures. Without a strong social consensus supporting the eradication of embedded sexism, such regulation can hardly be effective. If one seeks to enforce it thoroughly, one threatens to place an undue burden on women's bodies. Under such circumstances, the harm of allowing some unethical sex selection to go ahead may be outweighed by the need to protect women's procreative freedom and women's bodily autonomy.

In this paper, I use Taiwan as a case study to illustrate this point. Although the government officially condemns any sex selection done for non-medical reasons, the boy-to-girl ratio of newborns was 110:100 in 2003, much higher than the normal ratio of 105/106:100. Drawing on the already abundant literature that treats the morality of sex selection, this paper will focus mainly on the social implications of sex selection in Taiwan and on the justification for its regulation. To avoid the related complications stemming from abortion or the moral status of embryos, I limit this paper's discussion to pre-conceptive sex selections achieved through sperm-sorting technologies.

In the next section, I will briefly discuss the possible justifications for a ban on sex selections and provide a typology that distinguishes sex selections from one another on the basis of their different ethical weight. I will then present the case of Taiwan to argue that, because of the difficulties that attend both the policing of lines demarcating different types of sex selection and the negative effect of policing on women's bodily autonomy and procreative autonomy, the harm of a legal ban on sex selection is greater than the possible morality that such a law might maintain.

II. JUSTIFICATIONS FOR A BAN ON SEX SELECTION: A TYPOLOGY OF SEX SELECTIONS

In every society, reproduction has a fundamental importance for both the culture and the individual. It is through offspring that families are able to continue their lineage and preserve their wealth beyond each person's lifespan. Scholars have argued that reproductive choices are central to the personal meaning of one's life, to one's connection with future generations, and to the pleasures of child-rearing (Robertson, 1994).

Whether this right exists for sex selection is a more complicated issue. According to some people, if one feels that one cannot realize this choice unless the resulting offspring has a particular characteristic, a presumptive right to reproduction should also protect one's freedom to make such a selection (Robertson, 1994, pp. 152–153). Others maintain that this right must be protected, or else women might lose their way on the slippery slope of recently won reproductive autonomy (Warren, 1999, pp. 137–142).

But surely the freedoms associated with procreation are not without their limits. A society can limit presumptive rights if there is good and sufficient justification to do so. The significance of this point is that, in the case of pre-conceptive sex selection, even if sperm-sorting technologies do not involve abortion or the destruction of embryos, sex selection may remain morally controversial because of parents' different motivations for undertaking it.

This complexity requires us to examine with greater precision the justifications that correspond to the regulation of sex selection, as it stems from different motivations. People might select their baby's sex for several reasons. The reason often considered the most legitimate concerns parents who carry sex-linked genetic diseases that might be passed on to the offspring. These diseases include color blindness, hemophilia, and Duchenne's muscular dystrophy, just to name a few. In such a situation, it is generally agreed that sex selection is ethical and permissible. Even the countries that impose the strictest prohibitions on sex selection usually make an exception for these sex selections. I will call this *sex selection for medical reasons*.

But sex selection due to other motivations is, in terms of morality, more controversial. One of the motivations arises from boys' significance in the social context. For instance, boys can carry on the family's last name, provide much needed manpower on the farm, and bring home higher wages than girls can. I will call such a motivation *sex selection for social reasons*. Underlying this type of sex selection is generally a socially rooted stereotype of male and female identity.

Is all preference for the male sex or the female sex sexist and thus unethical? Some scholars seem to think so. It has been argued that selecting a child's sex occurs within the context of certain sexual stereotypes and, thus, constitutes sexual discrimination (Wertz & Fletcher, 1992, p. 249), regardless of the selection's actual impact on the relative frequency and status of baby boys and baby girls. Furthermore, it has been argued that because baby boys have long been privileged in the selection process, the sex ratio of the population will be upset and reinforce the social depiction of women as less worthy (Krugman, 1998; Farrell, 2002; Kim, 2000). Finally, because it is females who bear children and because recent medical technologies have made intrusions into women's bodies more possible, suggest that sex selection represents yet an additional oppression of women (Danis, 1995). Whereas the first argument points to fundamental moral concerns that will stand regardless of the social impact of sex selection, the latter two are consequentialist arguments whose validity depends on empirical evidence from each particular society and is hard to prove or disprove (Bubeck, 2002, pp. 218–220). I will save the empirically based arguments for later and analyze the morally based one first.

To begin with, the argument that equates sex selection with sexual discrimination, as righteous as it sounds, is over-inclusive; for, unless one states that to be free of sexual discrimination is – necessarily – to be absolutely gender-blind, one must accept that it is possible for people to prefer a particular sex without engaging in sexual discrimination, regardless of the sexism established in those people's society (Overall, 1987, pp. 23–27). Just as people may prefer a partner of a particular sex, parents may want to balance the sex of an unborn child with that of its sibling. This requires us to distinguish sex selections that are due to social reasons from those that are not. I will call this last type *sex selection for individual preference*. Hence, people's motivations for sex selection give rise to three kinds of sex selection (see Table 1 in the Appendix): people select their baby's sex for medical reasons (henceforth, known as A), social reasons (henceforth, known as B), and reasons of individual preference (henceforth, known as C), with the first classification typically found to be morally permissible, and the latter two in need of further analysis.

In fact, the line dividing the latter two types of motivation is more complicated than it at first appears to be. To begin with, parents who select their baby's sex for social reasons may have good excuses: parents who prefer a boy because he would fare better in their particular society might try to justify their action by equating it with their child's best interests. The parents might argue that to blame them would be to blame the victim, because as there is little that they can do about the sexism permeating their society, they can at least avoid making their child a victim of it, or more accurately, avoid giving birth to a prospective victim. Mary Anne Warren (1992, p. 234) argues that people who condemn this type of sex selection risk blaming victims for a crime beyond anyone's control. An objection to sex selections that are based on an embrace of sexism (which will be referred to as B1) should not lead us to ban sex selections that constitute a passive reaction against sexism (which will be referred to as B2). Likewise, among people who select their baby's sex for reasons of individual preference (C), some might prefer, on the basis of sexual stereotypes, a male to a female, or vice versa. An individual preference that is, nevertheless, informed by social norms (henceforth, known as B3) can be contrasted with pure individual preference (C).

All this discussion has revealed a subgroup composed of B1, B2, and B3 among the sex selections motivated by social reasons, with B2 morally distinguishable from B1 and B3. Moreover, this typology also distinguishes sex selection for reasons of socially inspired individual preference from sex selection for reasons of pure individual preference (C), which is morally neutral compared to other forms of sex selection (see Table 1 in the appendix).

Hence, although sex selection that is motivated by sexual discrimination should be banned, a ban covering all sex selections that are due to non-medical reasons requires further moral justification and risks being over-inclusive. And I think that, if B2 and C are objectionable, the problem lies not in the action itself but in the impact it may have in a particular society. Therefore, while sex selection in the case of B1 and B3 should be banned altogether, whether one should ban sex selection in the case of B2 or C requires further justification.

One of the justifications is that, although sex selections *per se* are morally neutral in the situation of B2 and C, they might have an adverse social impact. This objection gives rise to the consequential arguments against sex selection that I mentioned earlier. Namely, sex selection causes an imbalance in populations' sex ratio, reinforces women's secondary position as the less worthy sex, and causes unnecessary intrusions into women's bodies.

Yet, also in a consequential manner, Mary Anne Warren (1985, pp. 173–175) argued that sex selection can improve children's quality of life by ensuring that they are "wanted." This feeling, in turn, will enhance the balance of the family's life and thus lessen the mother's burden to give birth to more children in order to have a baby of the "right" sex (Warren, 1992, pp. 243–244). Also arguing for the merit of sex selection, others have argued that sex selection limits population growth because unwanted daughters will not be born before the wanted son is born (Singer & Wells, 1985, pp. 153–54).

But these consequential arguments are hard to prove or disprove, because they require empirical evidence from particular societies. As for the consequentialist argument that sex selection may increase unnecessary intrusions into women's bodies, since sperm-sorting technology usually requires only an injection of the "right" sperm into a woman's body, is a relatively less intrusive procedure compared to other types of sex selection.

Hence, *as long as the process requires women's informed consent*, I think that whether these pains are worthwhile is within the realm of freedom of reproduction and bodily autonomy and should therefore be up to women. The real danger of such a choice hinges on whether women can make a free and informed decision concerning reproduction and their body. I think that, in a patriarchal society, there is a real danger that women may not have enough freedom to make that decision. I will further discuss this point in the latter part of this paper.

To sum up, I have referred to different motivations for sex selection in order to present a typology wherein sex selections can be, in terms of morality, differentiated into four types: sex selection motivated by medical reasons, sexism, passive reaction against sexism, and pure individual preference. Because sex selections for reasons of either pure individual preference or passive reaction *per se* are morally neutral, I argue that a total ban on all sex selections that are based on non-medical reasons is morally unjustifiable if the ban lacks sufficient empirical evidence that such sex selections have an adverse social impact in that particular society.

But issues of sex selection do not end here. Rather, problems begin when the state intends to turn these moral judgments into public regulation; for, even if the government intends to draw the line only between sex selections that are based on medical reasons and those that are not, the lines can be hard to police and may prove to be more costly should the government over-intrude into people's reproductive choices. This complexity, relative to the case of Taiwan, is the focus of the following section.

III. SEX SELECTION IN TAIWAN

Under natural conditions, the ratio of boys to girls in newborns is 105/106:100. In Taiwan, however, the ratio rose beyond this number in the mid-1970s and, after 1985 (the year after the legislature relaxed its controls on abortion), steadily climbed to 110.3 (Ministry of Interior, 2003). In recent years, the number has remained stationary, only slightly fluctuating between 108 and 110 (Ministry of Interior, 2003).

The boy-to-girl ratio has also grown steadily for families that have more children. In the year 2000, the ratio of boys to girls in newborns was 107:100 for the firstborn, 108:100 for the second child, 119:100 for the third, and 135:100 for the fourth (Executive Yuan of the ROC, 2000). Compared to the fact that the average number of children in each family has declined from 1.76 in 1993 to 1.40 in 2001 (Ministry of Interior, 2003), many people appear to have borne a third or a fourth child simply to have a son.

This preference for a baby boy is also reflected in babies born by assisted reproductive technology (ART, as mentioned above). The official report on assistive reproductive technologies that are conducted in Taiwan shows that, from 1998 to 2001, the average sex ratio of newborns born as a result of IVF is 114, much higher than the average of 109 for children born without the aid of ART during the same period (Bureau of Health Promotion, 2003, p. 14).

However, abortion is not the only practice in Taiwan used for sex selection. Sex selection practices that do not involve any high-tech applications are available, as well. Among them are Chinese herbal medicines and special diets that influence hormones in the body. In addition, sex selection techniques such as sperm-sorting services are commonly available in Taiwan (Wu, 2002a). In fact, since entrepreneur-physician Dr. Ericsson introduced it on the island in 1981, sperm-sorting technology has gained widespread interest there among prospective parents seeking to bear a boy. Even though Ericsson authorized only four physicians to perform the technique when he first came to Taiwan, today similar methods of sperm-sorting technologies are commonly available in local Taiwanese clinics (Wu, 2002a, p. 1).

Apprehensive about the imbalanced ratio of baby boys to baby girls, the Department of Health has publicly forbidden the medical profession from providing sex selection services for non-medical reasons. Yet, the imbalanced ratio remains.

How can one explain the government's failure to ban sex selection in Taiwan? By examining the embedded sexism in the formal and the informal social structures of Taiwan, I will suggest that parents have good reasons to select the sex of their child. In these circumstances, it will be hard to distinguish selections of a child's biological sex that are based on ethical reasons from those selections that are not. Moreover, in the absence of a democratic consensus against it, a total ban on sex selections except those for medical reasons will unduly burden women's bodily autonomy, which deserves more protection in a patriarchal society.

IV. EMBEDDED SEXISM: THE SOCIAL MEANING OF HAVING A SON IN TAIWANESE SOCIETY

At first glance, one would think that the era of sexual inequality in Taiwan has passed into history and that sexism has no significant influence over sex selection. Women's participation in politics is fairly common. At the time of this writing, Taiwan has a female vice president and a female vice premier, and women elected to the legislature constitute 28.4% of the legislators (Executive Yuan of the ROC, 2002, p. 294). Women have also fared well in Taiwan's education system. Women constitute up to 50.3% of higher education students, 0.6% more than men in 2002 (Executive Yuan of the ROC, 2002, p. 286).

Underlying recent legal reforms has been the aim to eradicate sexual discrimination hidden in the law. For instance, the legislature amended the Civil Code three times since 1985 to give women an equal footing in decisions concerning the location of their home and their last name in marriage, in the exercise of parental rights over their children, and in their entitlement to half of their husband's assets earned during marriage should the couple be divorced or the husband die in the marriage. The legislature also passed the Family Violence Prevention Act to protect women and children from domestic abuse. In addition, the new Equality in Work Act addresses the issue of sexual harassment at work while providing equal pay, necessary maternity leaves, and other measures to make the work place friendlier for women.

But underneath the formal structure of progressive legal amendments, sexism is deeply entrenched in Taiwan's informal social structures. To begin with, even though women are no less educated than men, and even though the law guarantees equal pay for equal work, only 46.6% of women participated in the labor force in 2002, as opposed to 70.6% of men (Executive Yuan of the ROC, 2002, p. 285). Moreover, the average monthly income for women was only 35,868 NT dollars as recently as 2002, or 78.01% of men's average income that same year (Executive Yuan of the ROC, 2002, p. 287).

Women also suffer from a lower status not only in the workforce but also in more private settings, within the family, where women's secondary position in the patriarchal society is no less evident. Chen (2003, pp. 111–114) argues that because a daughter lacks permanent family membership in her natal family before she is married, and in her husband's family if she is divorced, parents tend to prefer a son who "stays in the family" and to bestow more resources on him. Acting in accordance with this privilege, parents and, indeed, the wider Chinese society also expect a son to carry out his duties to support his parents when they are old, to make offerings to them when they are dead, and to worship all the family's ancestors routinely. According to Chen (2003, pp. 121–122), traditional criminal law punishes only a son for failure to support his parents in their old age. Even though this is no longer the case in Taiwan's modern law, the custom persists in the form of a widespread social attitude. A study in 1994 shows that, while 39% of married sons lived with their parents and supported them financially, only 4% of married daughters lived with their parents (Lee, 1994, p. 1010). It has also been established that daughters tend to take care of their in-laws, while sons—with the help of their wives—take care of the sons' parents (Chen, 2003, pp. 136–137)

Even though modern legal amendments can change formal social structures and gear them toward gender equality, sexism is most entrenched in a traditional custom of worship that is beyond the legal realm. A son's privileged position confers on him the obligation to make offerings to his family's ancestors. In Chinese culture, the customary obligation to make offerings centers on both the idea of the afterlife and a respect for the deceased. Influenced by Buddhist and Taoist traditions, Chinese believe that there is an afterlife for the deceased. Therefore, Chinese usually maintain a small altar with a tablet that bears the last name of the family in their household. In front of the altar, the family makes routine offerings to ensure that their ancestors' needs are met in the afterlife. Because only sons bearing the family's last name can perform the duty of making offerings, those families that do not have sons bearing their name either adopt a son for the purpose or have their son-in-law "marry into" the family, hoping that their daughters will bear a son with the right last name to do the job (Chen, 1990, p. 171).

The pressure of this cultural norm is heavy. Mencius, the great Chinese philosopher, made the following admonishment: of the three greatest offenses one can commit against one's parents and ancestors, having no son is the worst (Book of Mencius, Part I of Li-Lo, Ch. 26). According to popular cultural belief, those who have no offspring possessing the same last name and, thus, able to carry out the worshiping and the offerings are doomed to suffer from cold and hunger in their afterlife (Tai, 1987, pp. 6–7). Therefore, to have a son not only ensures parents' needs in their current lives and their afterlives but also fulfills the parents' duty to their own ancestors and so on down the family line. Because the deceased—from the other world—oversee the lives of their offspring, every living person faces psychological pressure if he or she dares to offend ancestors by not producing a son.

Under such cultural norms, it is crucial to have a son that bears the same last name to "ensure there is incense burning on the altar" (Tai, 1987, pp. 6–7). This stipulation is further reflected in the difficulties with which efforts to amend the corresponding family law have contended. According to the current Civil Code, children must bear their father's last name unless their mother has no brother in her family to bear the last name. This rule applies even when the father agrees to let his son bear his wife's last name. The fact that bearing one's father's last name is the norm, and that bearing a mother's last name is allowed only when the mother lacks a brother, is telling: the mother's right is recognized only to the extent that the mother's father needs someone to whom he can pass down his last name. Hence, even the exception made for the mother is just another means to fortify the patriarchal institution on the mother's side.

Compared to all other amendments that have equalized the parental rights of mothers and fathers, the law on children's last names has thus become the most entrenched manifestation of patriarchy in family law. Although feminist activists have advocated an amendment to the law so that children can bear their mother's last name (as long as the father and the mother reach an agreement, therein), opposition in the legislature and throughout society remains staunch.

The entrenchment of patriarchy manifests itself not only in how difficult it is to amend the law, but also in how difficult it is to implement such an amendment after its passage. According to traditional Chinese custom, only males are entitled to inherit the family's assets. Yet, in the atmosphere of the new republic that overthrew the feudal Ch'ing Dynasty, the lawmakers of the Civil Code of the Republic of China, at the very beginning of its enactment in 1930, decided to guarantee the equal rights of daughters in that daughters could inherit their family's assets just as their brothers had traditionally done.

However, the patriarchal norm of inheritance was deeply rooted in society, and, more than seventy years after the amendment, women are still forced to formally forsake their right of inheritance so that their brothers can inherit the family's estate. According to a recent survey, only 43% of Taiwanese agree that daughters' inheritance of parents' estates should equal the sons' inheritance, and 25% even think that the eldest son or the sons, taken together, should inherit the entire estate in question (United Daily News, May 6, 2002, p. 15). Another study estimated that less than 20% of daughters actually inherit all, or part of, their parents' estate (Ming Shen Daily, April 22, 2004, Sec. A2). Thus, even though the practice is receding as women gain more power in Taiwan's society, it is still quite common.

Within such formal and informal social structures, the birth of a baby boy is "non-chang," which is Chinese for "gaining a jade," whereas the birth of a baby girl is "non-wa," which is Chinese for "gaining a tile." To share the good news with friends and relatives, the parents distribute rice cakes with eggs or drumsticks if they give birth to a son, and distribute plain rice cakes if they gave birth to a daughter.

The case of Taiwan shows that sexism can be so entrenched in formal and informal social structures that parents, either because they believe in sexual stereotypes or because they want to protect their children from belonging to the second sex, have good reason to select their child's biological sex.

These circumstances not only complicate the government's efforts to determine whether parents select their children's sex because of sexism, passive reaction, or pure individual preference, but also weaken the justification for a ban on all sex selections except those based on medical reasons. For instance, I am a feminist who happens to enjoy the life of a woman in Taiwan. But I am also a university faculty member who is economically and intellectually self-sufficient. I was pressured neither to bear a son nor to avoid a daughter. I would much rather have a daughter, who can share my experiences, than a boy, who cannot have my experiences. I know that I will be just as happy even if my baby turns out to be a boy, and will try my best to accept him as he is, and try to enjoy his companionship. But am I sexist for wishing for a little girl? Perhaps deep in my mind, there is something sexist about my assumption that a daughter would appreciate my experiences as a woman and that I would not prefer the life of a boy. Or is there necessarily a difference between a boy and a girl? If not, is it thus sexual discrimination to wish for a daughter? Alternately, if there is a difference between boys and girls, but the difference is socially constructed, am I guilty of reinforcing that stereotype if I wish for a son? What if I specifically wish for a girl in a patriarchal society? On the other hand, if there is a difference between boys and girls, but the difference is not socially constructed, am I allowed to wish for a baby girl?

But this difficulty of probing the mind for intentions surfaces whenever a person decides to act on his or her perceptions and motivations. For who would admit to embracing sexism when selecting a baby's biological sex? How can any mechanism in a patriarchal society successfully screen out sexism in people's minds? Indeed, unless we ban sex selection under any condition, we face the daunting task of probing the minds of every parent who seeks to use the technologies in question. But this would be unjustifiable, since parents need access to sex selection to avoid sex-linked genetic diseases.

But one might argue that, because it is hard to police the lines between types of sex selection based on non-medical reasons that have different ethical implications, it is still justifiable to ban all types of sex selection except those based on medical reasons, because this ban condemns the sexism underlying sex selections. However, I will suggest in the following section that, without a social consensus supporting the eradication of embedded sexism, such regulation not only will be futile but also may harm women's bodily autonomy and women's procreative autonomy, which perhaps deserve even more protection than the female embryos that might have been created without sperm-sorting technologies.

V. CONSENSUS IN NEED: THE PROBLEMATIC LEGITIMACY OF THE BANNING OF SEX SELECTION IN TAIWAN

Compared with the blurred and difficult-to-police boundaries among types of sex selection based on social reasons, sex selection based on medical reasons provides a clearer boundary to police, in terms of both morality and administrative cost. Even the most restrictive countries usually allow sex selection based on medical reasons.

Because this sex selection usually improves the child's prospects for a healthy life, it is free from criticism of sexism, even if the embryo of a certain sex is excluded from the process of fertilization. Moreover, because the permissibility of sex selection depends on physicians' medical opinions rather than on parents' motivations, this type of sex selection is also easier to police in terms of administrative cost.

However, underlying this classification is the assumption that whether or not a decision for sex selection is medically based is a scientific question that is value-free. Only thus can these sex selections withstand ethical tests better than other types of sex selection. However, this is not the case for all medical judgments. After all, are the medical reasons in question limited to serious diseases? For instance, statistics show that boys are more susceptible to sex-linked genetic diseases because they do not have an extra X chromosome for backup. Can I, for this reason, prefer baby girls just to be on the safe side, even though the risk is very remote, that is, even though I don't have a family history of such risks?

In contrast, if the medical reasons provided for sex selection are limited only to serious diseases, what constitutes "serious"? A patient with hemophilia might face a higher risk of injuries or complications requiring blood transfusions, but with due care, the patient would have a reasonable chance to prosper as well as anyone else. Can parents select their children's sex to prevent hemophilia? Besides, should there be a distinction between illnesses that have cures and those that do not? For, to allow sex selection for medical reasons risks introducing eugenic discrimination against people with disabilities. Many patients suffering from severe diseases or severe disabilities may nonetheless lead a meaningful life aside from the disease or before the disease actually strikes them. When can the physician make a sound judgment that there is medical reason for sex selection?

Given the ethics-based controversies concerning sex selection, and especially those controversies that concern (1) different sex selections that are based on non-medical reasons and (2) pertinent differences between medical and non-medical sex selections, if the government chooses to allow only sex selection based on medical reasons, at least there should be a social consensus that unites society and its physicians behind a condemnation of sex selection as immoral; for, without a social consensus against sexism, regulations that ban sex selection are hard to police when sexism is so entrenched in the social and legal structure and when justifications for a ban on all, except one type of, sex selection are shaky.

Over the years, the imbalance of the ratio between newborn baby boys to girls in Taiwan manifests this point. The case of Taiwan provides another unique situation: although government officials have repeatedly and publicly condemned sex selection based on non-medical reasons,[5] no formal legislation has been erected against it. Thus, neither the public at large nor their representatives have had a chance to deliberate the morality of sex selection and to take a stand, even though sex selection seems to be so widespread a practice.

Indeed, if the Department of Health (hereinafter, DOH) wishes to ban sex selection, it can rely only on the general clause that punishes violations of the medical ethics spelled out in article 25 of the Physician Act. The absence of a clear and special legal mandate is significant. The public has had no means by which to debate and influence how ART should be used, nor has the public had any say as to who should benefit from it.

A lack of public consultation is characteristic of the DOH's regulations. That the DOH has traditionally been headed by physicians falls under the rationale that only physicians can regulate physicians. Because the peer review mechanism is weak in Taiwan's medical profession, the DOH is responsible for the regulation of all medical practices. When a matter requires expertise in medicine or involves medical ethics, the DOH convenes advisory committees chaired by high-powered physicians who attempt to discern the more nuanced contours of the matter. Before 1990, when martial law prevented the public from participating in public affairs, the DOH thus established its authority in the medical profession with its technical expertise and professional seniority. Since the democratization of Taiwanese society, and even though the DOH has opened the advisory committees to a few lay people, the traditional style of self-contained regulation has remained basically intact.

But without public debate, there is little awareness as to the ethical problems that might arise in sex selection. Nor is there any social consensus as to the legitimacy of a ban on sex selection. Therefore, even though the DOH has made its policy clear, judging from the constant abnormality of the ratio between boys and girls in Taiwan's newborns, one can see that the regulation is hardly effective, if at all.

Physicians openly acknowledge both their sympathy for patients who are suffering from the pressure to bear a son (Li, 2001, pp. 16–18) and their awareness of their peers' involvement in sex selection (United Daily News, January 8, 2003, Sec. A11). One study estimates that 70–80% of obstetrician/gynecologists' clinics in the southern part of Taiwan still offer sperm-sorting technology (United Daily News, January 9, 2003, Sec. A15). Another recent study shows that, when parents who had two daughters requested that physicians detect their fetus' sex, 62.6% of the physicians agreed to do so (Wang et al., 2004, Sec. 5).

But the DOH prohibits only sex selection that is based on non-medical reasons so that, if physicians are willing to sanction (outside the medical profession's self-regulatory mechanisms) their patients requests for sex selection, then the DOH will have difficulty policing its own ban.

VI. THE NEED TO PROTECT ACTUAL WOMEN
OVER POSSIBLE EMBRYOS

Given that sexism permeates formal and informal social structures, and in the absence of a democratic consensus in Taiwan, I am doubtful whether its government can regulate sex selection fairly and effectively. But this should not be a problem if there is good reason for the government to enforce tighter regulations. However, in this final section, I argue that, without a social consensus supporting the eradication of embedded sexism, such regulation not only will be futile, but also may harm women's bodily autonomy and women's procreative autonomy, both of which are actual and both of which deserve even more protection than possible females embryos, which is to say, embryos that might be created without sperm-sorting technologies.

Under the social pressure to bear sons, Taiwanese women have had to fight hard for their bodily autonomy and against their husband's will. The government began to protect women from domestic violence in 1998 when the legislature passed the Prevention of Domestic Violence Act. And only in 1999 was it acknowledged in the Criminal Code that rape involving husband and wife constitutes a crime should the wife wish to pursue the matter through article 229–1.

However, in terms of reproductive freedom, Taiwanese women enjoy relatively free access to abortion only when their husband consents. To understand the complexity behind this hard-won freedom, one must understand the legal history of the Genetic Health Act, the law that authorizes women's access to abortion.

Currently, article 288 of the Criminal Code prohibits pregnant women from aborting their fetus and permits a maximum sentence of six months for women who violate this law. Anyone who helps to abort a fetus faces a maximum sentence of two years. However, the legislature's passage of the Genetic Health Act(1984) provides six exceptions according to which women can legally request the termination of their pregnancy. These exceptions include medical reasons such as possible genetic disease or handicap on the fetuses' side or risk of life and health on the mothers'side, and social reasons such as pregnancy stemming from rape or incest. An exception that has generated the most discussion in recent years is the sixth exception, which allows pregnant women to request an abortion if "continuance of pregnancy or birth will affect her mental health or family life." With this broad exception, physicians can freely perform abortion services despite the threat of punishment.

It is intriguing to note that the major purpose of the Genetic Health Act has been to exonerate physicians from blame and thus to free them from the threat of criminal prosecution when they perform abortions that "promote the population's quality." Liu, Chung-tong (1996, p. 235) argues that Taiwan's population policy has always emphasized the quality of the population rather than the quantity of the population; therefore, the government's official policy has permitted abortion for medical reasons, provided that it is performed in public hospitals that have been licensed to perform abortions. However, women seeking abortions still rely on private clinics because they provide more privacy. This illegal market in abortions has led to the sentencing of many physicians. When the government began to draft the Genetic Health Act to promote the population's quality, Taiwan's medical association seized the opportunity and used it to exonerate physicians from blame and to shield them from future prosecution for having provided abortions (Liu, 1996, p. 235).

Feminist activists, of course, have supported the act. But conservatives have worried that passage of this act not only shows disrespect for fetuses but also might lead to promiscuity in sexual relations (Liu, 1996, p. 235). Eventually, it was the intention of Taiwan's authoritarian government to decrease the birth rate that settled the debate. Therefore, even though some activists raised concerns about the fetus' life, the Act's most influential supporters had three major concerns: to control the rising population, to shield physicians from prosecution, and to enhance the quality of the population (Liu, 1996, p. 235).

The resulting Genetic Health Act is a mixed blessing for Taiwanese women's bodily autonomy. Article 9 of the Act permits abortion when there is concern for the health or the life of the mother or the fetus. It also allows women to abort their fetus if a continuation of the pregnancy would adversely affect their psychology or family life. This last exception gives women wide autonomy over their bodies. But to get an abortion according to this very exception, however, requires consent from the husband. Thus, for married women, the act grants the husband even more power over his wife and becomes another form of oppression over her body.

Studies suggest that, under such an abortion policy, parents may have achieved sex selection though a combination of ever developing prenatal tests and abortions (Wu, 1999, pp. 94–96; Wu, 2002b). A study shows that in 1991, fetuses that underwent Chorionic Villi Sampling (CVS, as mentioned earlier, which is a prenatal test that can determine the fetus' biological sex at 10 weeks of pregnancy) were 9 times more likely to have limb reduction defects than were fetuses that did not receive CVS (Ho, 2001, p. 10). What is telling is that all of these defects occurred in baby boys, and this finding has raised the suspicion that no baby girls were being born with such a defect because they were being aborted after CVS determined their biological sex (Ho, 2001, p. 10). Even though the practice of CVS lessened after the foregoing study was published, a study shows that researchers have continued to find newer technologies to serve the very same purpose (Ho, 2001, p. 10).

Given this social backdrop, amoral medical professionals have become willing accomplices in Taiwanese parents' desire to bear sons. Hospitals routinely inform parents of their fetuses' sex, without asking whether the parents wish to know. For a profit, physicians provide sperm-sorting services to desperate parents even though the effect of the technology is dubious.

With the strong social pressure in Taiwan for women to bear a son, there is a danger that a stricter ban on sex selection might force women either to rely on abortion to select their children's sex or to bear more children than they wish to. Either outcome will undermine both the justification for the enforcement of a stricter ban on sex selection—a justification that is associated with the protection of female embryos that might be fertilized—and the symbolic gesture to condemn sexism, only at the cost of rights that affect actual people: women's bodily autonomy and women's procreative freedom.

VII. CONCLUSION

Two factors render the regulation of sex selection extremely difficult. First, this regulation targets people's motivations for sex selections. As I have discussed in my typology of sex selections, the differentiating lines can be very treacherous to police and amount to the daunting task of mind-probing. Second, as a medical technology, sex selection has prompted the government to rely on the willingness of physicians to abide by the law. However, when there is no democratic consensus supporting the eradication of sex selection, such regulation can hardly be effective. If one seeks to enforce it thoroughly, one threatens to place an undue burden on women's bodies. Under such circumstances, the harm of allowing some unethical sex selection to go ahead may be outweighed by the need to protect women's procreative freedom and women's bodily autonomy.

Therefore, I contend that the morality of sex selection, as an issue open for exploration, merits more informed attention and more public debate because democratic law should not transcend society by evading either public debate or social consensus, especially when enforcement of the law is so deeply entrenched in a society's formal and informal social structures. Rather than limit people's freedom to use technology to realize their reproductive plans, it would ultimately be more justifiable and effective to change sexist social structures that induce people to prefer sons.

APPENDIX

Table 1. Typology of Different Motivations for Sex Selection

Motivation related to the context of a sexist society	Medical reasons (A)	Social reasons (B)	Reasons of individual preferences (C)
Exist even in non-sexist societies	A		C
Exist only in sexist societies		B1: Support for sexism B2: Passive reaction against sexism	B3: Personal preference based on sexual stereotypes

NOTES

[1] This paper benefited from the research grant from Taiwan's National Science Council "In Search of an Institutional Paradigm for Genetic Technology: An Analysis of Genetic Engineering and Reproductive Technology (NSC91-3112-H-033-001)." An earlier version of this paper was presented at the 50th Annual Meeting of the Law and Society Association, Chicago, U.S.A., May 30, 2004, and the Fourth International Conference of Bioethics: Biotechnology, Family, and Community, Chung Li, Taiwan, June 30, 2004. The author wishes to thank Chia-Ling Wu, Margaret Sleeboom, John Harris, and H. Tristram Engelhardt, Jr., for their generous offers of helpful information and comments.

[2] Article 14 of the Convention for the Protection of the Human Rights and Dignity of the Human Being with Regard to the Application of Biology and Medicine: Convention on Human Rights and Biomedicine, Oviedo, 4.IV. 1997. The Convention was adopted by the Committee of Ministers on November 19, 1996, was opened for signature April 4, 1997, and has been in force since December 1, 1999. For the text of the convention, visit http://conventions.coe.int/treaty/en/treaties/html/164.htm (Accessed on February 14, 2005).

[3] As of February 16, 2005, of the 46 members of the Council of Europe, 19 member states including France, Italy, Spain, Denmark, Norway, Sweden, and Russia have signed the Convention, and among these 19, 13 ratified the Convention. However, member states such as the United Kingdom that did

not sign it, have domestic regulations that ban sex selection. For the status of the Convention, see http://conventions.coe.int/Treaty/Commun/ChercheSig.asp?NT=164&CM=8&DF=2/16/05&CL=ENG (Accessed on February 16, 2005).

[4] The Human Fertilisation and Embryology Authority recommended, on November 12, 2003, that sex selection techniques involving sperm sorting should be regulated in the United Kingdom and that the current policy of allowing sex selection only to avoid serious sex-linked disorders should continue. For the Press Release, visit http://www.hfea.gov.uk/PressOffice/Archive/1068631271 (Accessed on February 14, 2005).

[5] The Department of Health issued a notice to clinics, warning them not to perform sex selection services more than once. See, for example, Ming Shen Daily News, p. 21, June 5, 1995; United Evening News, p. 2, January 26, 2001.

REFERENCES

Björndahl, L., & Barratt, C. (2002). 'Sex Selection: A Survey of Laboratory Methods and Clinical Results,' Survey prepared for the Public Consultation of Human Fertilisation and Embryology Authority.

Bubeck, Diemut (2002). 'Sex Selection: The Feminist Response,' in J. Burley & J. Harris (Eds.), *A Companion to Genetics* (pp. 216–229), Blackwell Publishers, Oxford, U.K.

Bureau of Health Promotion, Department of Health, Executive Yuan of the ROC. (2003). 'Analysis Report on Assisted Reproduction Technology from 1998 to 2001 in Taiwan' [On-line, In Chinese]. Available: http://www.doh.gov.tw.

Chen, C. (1990). *'Fang' and the Traditional Chinese Family System* [In Chinese]. Lian-Ching Publishing Co, Taipei.

Chen, C. (2003). Gendered Lives under Neutral Laws: Women, the Family, and Feminist Legal Reform in Taiwan. S.J.D. dissertation, University of Michigan Law School.

Danis, J. (1995). 'Sexism and "the Superfluous Female": Arguments for Regulating Pre-Implantation Sex Selection,' *Harvard Women's Law Journal*, 18: 219.

Executive Yuan of the Republic of China. (2000). *ROC Yearbook 2000* [On-line]. Available: http:www.roc-taiwan.org/taiwan/5-gp/yearbook /chpt02-1.htm.

Executive Yuan of the Republic of China. (2002). *Social Statistics Index* [On-line]. Available: http://www.dgbas.gov.tw/dgbas03/bs2/92chy/table/a203.xls.

Farrell, K. (2002). 'Where Have All the Young Girls Gone? Preconception Gender Selection in India and the United States,' *Indiana International and Comparative Law Review*, 6, 253–281.

Ho, H. (2001). 'Taiwan Women's Mission to Bear Sons,' *Newsletter of Applied Ethics*, 17, 9–12.

Holmes, H.B. (1995). 'Choosing Children's Sex: Challenge to Feminist Ethics,' in J.C. Callahan (Ed.), *Reproduction, Ethics, and the Law: Feminist Perspectives* (pp. 148–177), Indiana University Press, Bloomington.

Jones, O.D. (1992). 'Sex Selection: Regulating Technology Enabling the Predetermination of a Child's Gender,' *Harvard Journal of Law & Technology*, 6, 1–62.

Kim, N. (2000). 'Breaking Free from Patriarchy: A Comparative Study of Sex Selection Abortions in Korea and the United States,' *UCLA Pacific Basin Law Journal*, 17, 301–325.

Krugman, A. (1998). 'Being Female Can Be Fatal: An Examination of India's Ban on Prenatal Gender Testing,' *Cardozo Journal of International & Comparative Law*, 6, 215–237.

Lee, Y., et al. (1994). 'Sons, Daughters, and Intergenerational Support in Taiwan,' *American Journal of Sociology*, 99, 1010–1041.

Liu, C. (1996). 'Section on Health and Reproduction,' in Y. Liu (Ed.), *1995 Whitepaper on Women's Rights* [In Chinese] (pp. 219–254), Shih Pao Wen Hua Publishing, Taipei.

Mencius, 'Li Lo,' in P.Y. Hsieh, et.al. (Eds.), *Readings from the Four Classics* [In Chinese] (pp.4458–4512), San-Ming, Taipei.

Ming Shen Daily, April 22, 2004, Sec. A2.

Ministry of Interior, Executive Yuan of the Republic of China. (2003). *Annual Statistics Report* [On-line]. Available: www.moi.gov.tw/w3/stat/home.asp.

Overall, C. (1987). *Ethics and Human Reproduction: A Feminist Analysis*, Allen & Unwin, Boston.

Robertson, J. (1986). 'Embryos, Families, and Procreative Liberty: The Legal Structure of the New Reproduction,' *Southern California Law Review*, 59, 939–1041.

Robertson, J. (1994). *Child of Choice*, Princeton University Press, New Jersey.

Singer, P. & Wells, D. (1985). *Making Babies: The New Science and Ethics of Conception*, Scribners, New York.

Sumner, A., et al. (1972). 'Distinguishing Between X, Y, and YY-bearing Human Spermatozoa by Fluorescence and DNA Content,' *New Nature Biology*, 229: 231.

Tai, Y., & Tai, T. (1987). *Chinese Law of Inheritance* [In Chinese], San-Ming, Taipei.

United Daily News, May 6, 2002, p. 15.

United Daily News, January 8, 2003, Sec. A11.

United Daily News, January 9, 2003, Sec. A15.

Wang, M., et al. (2004). 'Comparison of Attitudes toward Genetic Testing among Health Care Professionals and Nonprofessionals' [In Chinese], paper presented at the Fourth International Conference of Bioethics, organized by the Graduate Institute of Philosophy, Chung-Li, Taiwan.

Warren, M. (1985). *Gendercide: The Implications of Sex Selection*, Rowman & Allanheld, New Jersey.

Warren, M. (1992). 'The Ethics of Sex Preselection,' in K. Alpern (Ed.), *The Ethics of Reproductive Technology* (pp. 137–142), Oxford University Press, New York.

Warren, M.A. (1999). 'Sex Selection: Individual Choice or Cultural Coercion?' in H. Kuhse & P. Singer (Eds.) *Bioethics: An Anthology* (pp. 137–143), Blackwell Publishers, Oxford, U.K.

Wertz, D. & Fletcher, J. (1992). 'Sex Selection through Prenatal Diagnosis: A Feminist Critique,' in H. Holmes & L. Prudy (Eds.), *Feminist Perspective in Medical Ethics* (pp. 240–253), Indiana University Press, Bloomington.

Wu, C. (2000). 'Autonomy in Reproduction,' in Awaking Foundation (Ed.), *1999 Women's Rights Report* [In Chinese], (pp. 93–103). Taipei: Awakening Foundation.

Wu, C. (2002a). 'Why Hasn't Sperm-Sorting Technology Become Controversial in Taiwan?' [In Chinese], paper presented at the Workshop of Gender and Health, December 7, 2002, National Yang Ming University, Taipei.

Wu, C. (2002b). 'Pu Tong Te Hsing Pieh, Pu Tong Te Cheng I?' ['Different Sex Selections; Different Issues?'], *Newsletter of Practical Ethics*, 17 [In Chinese], 19–24.

CHAPTER 9

MODERN BIOTECHNOLOGY
AND THE POSTMODERN FAMILY

LEONARDO D. DE CASTRO
University of the Philippines, Philippines

I. TWO TYPES OF CHALLENGE TO HUMAN RELATEDNESS

This paper is about the consequences of challenges posed to traditional conceptions of the family by developments in modern medicine. It distinguishes two types of challenges arising from the options that have been made possible by advances in the application of modern methods to the practice of medicine. The first type of challenge lies in extending the frontiers of human freedom.The second type of challenge has to do with extending the frontiers of human responsibility. Much attention has been paid to the challenge to extend the frontiers of human freedom. Now, we should realize that the challenge to extend the frontiers of human responsibility is equally deserving of a place in our collective consciousness. Case studies of organ transplantation confront us with specific issues pertaining to our notion of human responsibility. They indicate a need to expand the concept of human interrelatedness and, in this connection, the concept of the family.

This paper also makes an assessment of the implications of these challenges for Confucian and Western conceptions of the family. The observation is made that differences between opposing conceptions of the family are tending to be blurred by technological advances. Although families constructed along Confucian and Western paradigms are approaching the challenges differently, the overall outcome represents a diminished heterogeneity reminiscent of a globalizing cultural landscape. Insofar as the recruitment of donors for transplantable organs are concerned, "Western" families are acquiring Confucian characteristics while Confucian families are occasionally finding it useful to invoke Western values in order to protect the interests of individual members.

The obvious challenges to traditional conceptions of the family lie in the practice of technologically assisted reproduction and, prospectively, in reproductive cloning. The possibilities arising from the use of various techniques have generated a lot of controversy. Other challenges to traditional conceptions of the family appear to have drawn minimal attention. As a result, we have overlooked important lessons

S.C. Lee (ed.), The Family, Medical Decision-Making, and Biotechnology, 113–125.
© 2007 *Springer.*

about human relatedness. These lessons can go a long way in helping the public understand the impact of modern biotechnology on the family and cope with the controversies that arise.

Developments in the area of organ transplantation illustrate emerging concepts of relatedness that pose unprecedented challenges to the concept of the family. These challenges may be attributed not only to advances in biotechnology but also to structural changes in society brought about by greater interdependence among non-relatives and genetic or personal "strangers". Given the imposing developments in biomedicine, we have to accept that these challenges are giving rise not only to controversies but also to the imminent reconceptualization and realignment of human families on a global scale.

For this paper, certain assumptions about the differences between Confucian and Western conceptions of the family are made. A Confucian family is taken to be one that is hierarchical in authority structure. Even when there is an occasion for consensual decision-making, the hierarchical chain is able to assert itself. The contrast is with Western horizontal kinship characterized by the recognition of broad individual prerogatives capable of justifiably overriding consensual decisions or long held values. Moreover, membership of a Confucian family is presumed to be extended, with second or third degree relatives – not necessarily related by blood – joining the inner circle of attention, socialization, and decision-making. On the other hand, a Western family is typically small, with a scope that is confined to the most immediate members. Within this small domain, the authority is still diffused, with individual members being seen to enjoy a high degree of autonomy in decision-making.

Notwithstanding this putative distinction between Confucian and Western conceptions, it has to be clarified that these are not clearly geographic categorizations. There is no clear line that can be drawn across the globe to demarcate Confucian societies from Western societies insofar as family conceptions are concerned. Many families living in the West identify themselves with Confucian traditions and thus are characteristically Confucian. At the same time, many families living in the Confucian societies identify with Western Traditions and are thus, characteristically Western. Even the origins of particular families are not definitive indicators of the way that their thinking can be characterized.

II. REPRODUCTIVE REVOLUTION: EXTENDING THE FRONTIERS OF HUMAN FREEDOM

The successful use of in vitro fertilization for human reproduction has served as a monument to an era in which new options for making babies have been realized. These options have amounted to challenges to extend the frontiers of human freedom, revise the boundaries between families, and redefine human relatedness. For example, assisted reproductive techniques have made it possible for an unmarried virgin to become a genetic mother, for a woman to give birth to a granddaughter by serving as surrogate to her daughter, or for a lesbian mother to give birth to a genetically related child without getting involved in a hetero-sexual relationship. If human reproductive cloning eventually becomes successful, women could have the freedom to raise children without need for human sperm and, therefore, without any contribution at all by a male to the reproductive process.

These possibilities have been the subject of discussion and controversy for some time. This paper will not go into the issues involved. The examples have merely been given to illustrate the massive enhancement of options for human reproduction. What were previously thought to be natural limits to human freedom have been breached and continue to be breached. The revolutionary enhancement of repro-ductive options has resulted in the reinvention of the concept of the human family. Some people have found the extension of human freedom and the reinvention of the family to be threatening to human values and therefore unwarranted. On the other hand, others have argued that the developments represent an exercise of human freedom that can only be suppressed if there are imminent dangers to human beings or to their quality of life that need to be avoided.

A challenge to extend the frontiers of human freedom arises when developments in medicine make it possible for human beings to exercise options that benefit themselves. The options may be exercised to promote their welfare or interests. The availability of in vitro fertilization has given couples or individuals reproductive options for their own benefit. The options have posed a challenge to them to extend their own freedom. The challenge to extend one's freedom through the use of modern reproductive technologies involves, in most instances, a correlative challenge to reconceptualize and redefine the family, whether from a Confucian or from a Western perspective.

III. ORGAN TRANSPLANTATION: EXTENDING THE FRONTIERS OF HUMAN RESPONSIBILITY

This paper focuses not on the challenge to extend freedoms but on another type of challenge – the challenge to extend the frontiers of human responsibility. A challenge to extend the frontiers of human responsibility arises when developments in medicine make it possible for human beings to take up certain options that benefit others rather than themselves.

An example of this type of challenge is illustrated in the developments surrounding the transplantation of organs. National authorities are now being forced to rethink policies that frown upon the recruitment of donors who are not geneti-cally related to the possible recipients of their organs. Many countries are having to deal with existing policies restricting living donor transplantation to those who are related.

One focus of attention concerns a restriction to the effect that living donors could only donate an organ to someone with whom they are related by consanguinity. Among the reasons given for this restriction is the expectation that it serves to minimize the coercive pressures that could be brought to bear on non-relatives, especially those that are not attributable to altruistic motivations. There seems to be also a prevailing view that, given the nature of the risks involved, non-relatives could not safely be presumed to be able fully to understand the implications of going through the process of donating an organ. Their consent to donate cannot be ascertained to be truly free and informed. Thus, when non-related living organ donation has been permitted, it has been subjected to tight safeguards in terms of regulations or legislation. In many cases, it has also been subjected to the prior review of hospital ethics committees. In view of advanced medical technology and pressing insufficiency of donor organs, these policies are being questioned more widely.

As a consequence of advanced medical-surgical techniques and drug development, organ transplantation is becoming safer for recipients and donors alike. Powerful drugs are making rejection a minimal issue, helping to render obsolete a primary hurdle to non-related organ transplantation. To illustrate:

> The current results from more than 200 U.S. transplant centers demonstrate that kidney grafts from living unrelated donors continue to have excellent long-term survival rates despite a highdegree of HLA incompatibility ... [and] living unrelated donors exhibited short- and long-term graft outcomes similar to values of sibling donor transplants. At one-year post-transplantation, all living donor types exhibited significantly improved adjusted graft survival rates compared with cadaveric kidney transplants; however, the five year gs [graft survival] of living unrelated and sibling donor transplants continued to be good, but the parent donor recipients fared much worse and, in fact, had long-term survival rates similar to cadaver kidney transplants (Gjertson & Cecka, 2000).

Research findings such as these have put tremendous pressure on societies to allow non-related living organ donation. After all, more living donor organs for transplant from any source means longer or higher quality lives for more people in general.

Also being challenged now are policies encouraging preference for the procurement of cadaver organs rather than living donor organs. Studies show that patients transplanted with organs from living donors have had higher survival rates compared to those transplanted with organs from cadavers. For example, figures from the Organ Procurement and Transplantation Network in the United States indicate that patients receiving kidneys from living donors have higher survival rates than patients receiving kidneys from cadavers. The 5-year survival rate for transplants performed between 1995 and 2002 was at 90.5% for living donors compared to 82.5% for cadaveric donors.[1] For liver transplants, the 5-year survival rate was at 80.0% for living donors compared to 72.1% for cadaveric donors.[2]

Graft survival rates are also significantly better overall when the source is a living donor than when the source is a cadaver. For transplants performed between 1995 through 2002, the 5-year survival rate for kidney transplants was at 78.6% for living donors compared to 65.5% for cadaveric donors.[3] For liver transplants, the 5-year survival rate was at 70.8% for living donors compared to 64.4% for cadaveric donors.[4] These comparative survival rates explain why doctors and patients are encouraged to pursue living rather than cadaveric donors. Since 2001, the annual number of living donors of kidneys has exceeded the number of cadaver donors in the United States.[5] Overall trends in Europe still show a predominance of cadavers as sources of transplanted organs, but higher rates for living donor transplants have already been experienced in Cyprus, Greece, Israel and Romania (Council of Europe, 2004b). In South America, countries such as Brazil, Costa Rica and El Salvador have had higher kidney transplant rates from living than from cadaver donors (Council of Europe, 2004a).

IV. INCREASING HUMAN INTERDEPENDENCE IN THE AREA OF ORGAN TRANSPLANTATION

What is perhaps the single, most important message that can be gleamed from these statistics is the increasing degree of interdependence among human beings in matters of health in general and, in particular, in medical matters that can be addressed by recourse to organ transplantation. This interdependence lies in a number of interrelated aspects: (1) the improving potential to prolong life or to improve its quality through organ transplantation; (2) the perceived insufficiency of willing organ donors; (3) the existence of a huge gap between the essential need of human beings and the assistance that fellow human beings are able and willing to provide; (4) the inability of members of a genetic family to provide the essential needs of their relatives; (5) the availability of innovative ways to broaden the pool of human beings from whom assistance can be sought when an organ transplant is seen to be medically indicated; and (6) the increasing capability of those who are not genetically related to provide the essential health needs of fellow human beings by making an organ available for transplant.

The improving potential to prolong life or improve its quality through organ transplantation is highlighted, for example, in comparative studies that indicate that recipients enjoy a better quality of life than people who have undergone other therapies (Keown, 2001). There are also potential savings on expenses, as organ transplants present the best cost/benefit ratio (Miranda et al., 2003).

The perception of insufficiency of willing organ donors is supported by the long waiting lists for various transplantable organs in various countries throughout the world. People are dying while waiting to receive organs. The mortality rate for people on waiting lists ranges from 5–30% depending on country and type of organ (Miranda & Matesanz, 1998).

The long waiting lists are evidence of a wide gap between the number of patients who are waiting for other people who could help them and the number of people who are offering them organs for transplant. This is a gap that is increasingly requiring attention because advances in medicine and technology are making organ transplantation a treatment option for more and more people. However, this development has not been matched by a proportionate growth in the number of organ donations.

The inability of members of a genetic family to fill the essential health needs of their relatives is also manifested in these long waiting lists. The reason for a particular family member may be some kind of medical incompatibility, a disqualifying health condition, psychological or emotional difficulty, or plain and simple indifference. In some cases, the refusal has come from the prospective recipients themselves because of their unwillingness to involve their children or some other close relative in the concomitant risks. Whatever the reasons may be, the inability of close relatives to contribute transplantable organs has been partly responsible for making waiting lists longer. In one study, the failure of spouses to come to the rescue of the transplant patients was cited as a significant block to the shortening of waiting lists:

> Among the 43,000 patients waiting for a kidney transplant in the United States, as many as 6,000 potential spouse donors could be available.... If the 6,000 potential spouse donors became actual donors, the U.S. waiting list could be reduced by 15%.... Clearly, the 1,765 total accrued spouse transplants and the current rate of 374 spouse transplants per year fall far short of this potential (Gjertson & Cecka, 2000).

The effect of the inability or unwillingness of spouses is to shift the burden of responsibility to persons outside the family.

In response, various types of solutions are being offered to bridge the gap between the need for transplantable organs and the availability of donors. Innovative approaches are being pursued in an effort to expand the pool of human beings who can contribute to the satisfaction of the health needs of all. The following are some of the methods that various countries have been tried: (1) the use of monetary incentives; (2) the use of organs from executed prisoners; (3) the use of non-heartbeating donors; (4) moving the age limit in either direction for allowable donors; (5) a liberal policy towards living non-related donors; and (6) a system of "opting out" or presumed consent.

One thing that is characteristic of these methods as options in dealing with organ ailment is that they rely on the capability and willingness of human beings to make a direct contribution with their own organs. Ordinarily, the willingness to make a direct contribution with transplant organs is expected to be manifested by close relatives. Given the widespread public awareness of the options being made possible by advanced technology the call for contributions is now extending further beyond the confines of the recipient's genetic family.

V. ORGAN TRANSPLANTATION: A CHALLENGE TO GIVE MORE

The effort to respond to the health needs of another human being by making a direct contribution with one's own organs may be contrasted with the effort to cure a disease or improve the quality of life that rests on the ability of researchers to discover new drugs, implements or medical procedures. In the latter, the challenge is for researchers to work more and gain more success in discovery and invention. In the former, the challenge is for people in general to give more — of their own body. In the end, even the efforts of transplant coordinators to improve the system of recruitment or of transplant doctors and researchers to try to make the procedures safer and find better drugs translate into increased pressure on other human beings to make the transplant sacrifice for the sake of the patient in need. Such efforts could not meet with success unless there is a direct contribution of a transplantable organ. This is the bottom line.

Monetary incentives are meant to have the effect of enabling people to overcome fears or taboos that they may have about having an organ surgically removed. The use of organs from executed prisoners invokes (rightly or wrongly) the solidarity of human beings in the effort to justify the coercive means that are employed. The use of non-heartbeating donors, living non-related donors, and people who could otherwise be considered too old or too young to voluntarily and knowingly make the organ donation are all meant to broaden the base for the retrieval of organs for transplant.

What is happening then is that people who would not ordinarily have had any involvement in the problems of others somehow become involved because modern technology has made it possible for them to provide assistance. Because of modern technology, those who were previously distant and uninvolved have been provided a reason to be close and involved. People who would not have been regarded as having a responsibility for the treatment of particular patients in the past are now seeing the burden of helping to save or extend lives by organ transplantation passed on to them.

On the other hand, patients who previously thought their chances of getting a lease on life or a better quality of life rested exclusively on the availability of cadaver donors or living related donors have now been given a reason to look forward to contributions by non-relatives. With more effective drugs to combat rejection, the difference in success rates between transplants from related donors and those from non-related donors is being minimized. Hence, notwithstanding the genetic distance between pairs of non-related patients and donors, the parties are being drawn closer together by the essential health need of one party and the capacity of the other to fill that need. The genetic distance is being bridged by increased interdependence among unrelated persons in this area of critical care.

For the Western family, this development is something that involves a basic clash of values since the compelling nature of critical health care intrudes upon the autonomy of those who are called upon to render assistance or sacrifices. In a Western country like the United States, this clash of values was previously resolved in favor of autonomy by a legal prohibition of monetary incentives that was meant to preserve the capacity for autonomous decision-making of prospective donors.

However, the situation appears to be changing. Bills seeking to provide compensation to organ donors are in the congressional mill and are now thought to have better chances of being passed than before. The shift of dominance from cadaver to living donor kidney transplants may be seen as an indication of greater acceptance of an increased role for non-relatives since relatives alone would not suffice to provide the transplant needs of end-stage renal disease patients.

In Confucian societies, the culture of extended families has been more hospitable to living non-related organ transplantation. Since helpers, family drivers or other household hands can easily be considered part of the family; laws limiting organ donations to relatives appear to have had little effect.

VI. THE GENETIC FAMILY AS LOCUS OF RESPONSIBILITY

To be sure, even in the face of varying levels of attachment and closeness within a genetic family, it continues to be the locus of responsibility for the needs of its members. When there is a member in need of an organ transplant, there is pressure on parents, children or siblings to have themselves tested for compatibility and, if appropriate, to donate an organ. The pressure increases as they are told about the benefits for living donor recipients, the high probability of success, and the low risk of physical complications. They could also be induced by the anticipation of an improved sense of well-being and a boost in self-esteem arising from the perceived nobility of their purpose.

Human families have been rooted historically in genetic kinship. They are nurtured and sustained by the members' capacity and commitment to care for one another. Crucial to caring is the satisfaction of basic needs, including those that pertain to health: the preservation of life and the promotion of a satisfactory quality of life. The interdependence in the satisfaction of these needs has kept families together. It has been a key component in defining who belongs to a particular family or who does not belong to it. The interdependence covers social needs, economic needs, and physical safety or security. In times of need for these essential concerns, people tend to rely, first of all, on those to whom they are genetically related.

However, it could very well be that reliance in those to whom people are genetically related is occasioned not by the genetic relationship itself but by the physical accessibility of the relatives. It is one thing to say that we depend on our relatives because we are genetically related to them but another thing to say that we turn to our genetic relatives because of our emotional bonds or other factors that are merely incidental to their being related genetically to us.

In times of great and difficult need, the reliability of genetic relatives tends to be put to the test. This test arises quite readily when an organ transplant is recommended for a family member and the relatives, rightly or wrongly emerge as the immediate candidates for organ donation. Whether it is for medical or for other reasons, the closest relatives are expected to be the first to come to the rescue of the stricken family member. However, the response does not, and need not always conform to that expectation. Notwithstanding the experts' assurances regarding the relative safety of kidney donation, we continue to hear stories of siblings, parents or children being unable or straightforwardly unwilling to make the transplant sacrifice.

In this regard, one could say that there is a greater expectation in Confucian societies for relatives to make their organs available for donation. The hierarchical family structure invests the head with authority to encourage an eligible member to make the sacrifice. Even without explicit prodding, an eligible family member would feel the weight of expectation on her shoulders.

On the other hand, this burden of expectation is less felt in the West because the recognition of individual sovereignty and autonomy relieves one of a responsibility to assume the risks associated with organ donation in order to ensure the survival of a relative. This characterization is supported by policies seeking to protect the prospective individual donor from the pressures that may be exerted by those wanting to help a diseased brother or sister. For example, there is a policy of not disclosing that a prospective sibling donor is undergoing tests for eligibility or compatibility. After the tests are undertaken, the policy is not to disclose the results unless the sibling has decided to donate the needed organ. The idea behind the policy is to provide the proper atmosphere for the sibling to make a decision with the least pressure from the rest of the family. The presumption is that the freedom of the individual is more important than the beneficial consequences that the donation could bring about.

VII. LIMITS TO GENETICALLY BASED RESPONSIBILITY

The protection for prospective sibling organ donors serves to promote autonomous individual decision-making and prevent family expectations from giving rise to psychological bondage. Interdependence among those who are genetically related is not necessarily seen as an ideal to be cultivated in the face of essential health needs.

Indeed, even among the closest of relatives, organ donation is not necessarily done on a strictly altruistic basis. Offers to provide a transplantable organ are occasionally made with expectations of reciprocity. In the case described below, the donor obviously expected reciprocal compensation for the organ donation:

> Case 1: Anna and her immediate family had trouble making ends meet. After consulting with her parents and siblings, she decided to donate a kidney to her rich second-degree cousin, Betty. At first, Anna was afraid to go through the procedure. However, the hopeful anticipation of being rewarded enabled her to overcome her fears. After the transplant, Betty's family immediately offered a monetary gift to Anna and her family, which the latter accepted. The families were drawn closer together and Betty's family has always been ready to provide assistance to Anna's family for various kinds of essential expenses. The relationship between the families has gone on even though it has been 15 years since the transplant took place, and Betty died 3 years ago of a condition unrelated to the kidney transplant.

Aside from being an example of a donation that is not based purely on the recognition of a genetic bond, this case is illustrative of the close kinship that was more of an effect rather than a cause of the interdependence among family members. The mutual capacity to provide for the needs of each other brought the families together, to an extent that their genetic relationship could not, prior to the medical developments described.

This previous case makes for good contrast with the following:

> Case 2: The functioning of Mrs. Castillo's kidneys was deteriorating rapidly and she was getting distressed by their inability to find a transplant donor even after a year's search. Mrs. Castillo belonged to a rich family that employed several domestic helpers and a chauffer, all of whom lived in quarters at their family compound. While driving Mrs. Castillo to and from the hospital, the chauffer had a chance to listen to his employer express her fast-growing anxiety to her friends. He became aware of his employer's condition and the desperation that was starting to grip her. Without being asked, the chauffer decided to offer one of his own kidneys. The offer was accepted. The doctors successfully performed what they regarded as a transplant from an emotionally related donor.

This case illustrates a close kinship that has no genetic basis. Having been employed for a number of years and sharing the same residential compound with his employer, the chauffer sensed the urgency and responded to it. He went beyond merely recognizing the urgency to responding to it in a way that close relatives are the ones usually expected to do. In this case, the chauffer did not ask for a reward, though one may argue that he could have had that at the back of his mind or that he felt bound by some debt of gratitude to his employer. Real motivations are always difficult to determine but perhaps it is useful to look a bit more closely at how debts of gratitude contribute to non-genetic kinship, as illustrated in the following case:

> Case 3: When he was younger, Danilo was sometimes involved in violent confrontations with other people in his neighborhood. After one such confrontation, Danilo was on the verge of being arrested by the police. Efren, who was then an elected village leader, interceded in Danilo's behalf and convinced the arresting officers to let him go. He was released in Efren's custody.

Ten years later, the Danilo still vividly remembered his long-standing debt of gratitude to the already retired village official. When he heard that Efren needed a kidney, he offered to be a donor. According to him, that was the perfect way to show his appreciation for the help that he once received from the village official.

A number of arguments can be raised in favor of, or against the donation of transplantable organs under the circumstances described. For example, some people feel that having a debt of gratitude can be so overwhelming as to constitute coercive influence, preventing the agent from acting or deciding autonomously. The same thing can be said about the conditions of employment that could have unduly pressured the chauffer in Case 2 to decide to offer one of his kidneys.

VIII. RECIPROCITY AND RESPONSIBILITY: TRANSCENDING GENETIC BOUNDARIES

While precautions are thought to be necessary to ensure that, say, siblings are not unduly being coerced to make a transplantable organ available, people do generally take for granted that the pressures weighing on them constitute a positive inducement associated with the feeling of belonging and oneness among family members. These pressures help to promote a sense of responsibility and to generate a feeling of heroic satisfaction in the related donor. If this is how we look at the commitments within genetic relationships, there seem to be no reason why we could not look similarly at commitments arising from primarily emotive relationships that have no genetic roots. The chauffeur's offer in Case 2 could be seen to result from a positive inducement that is associated with a feeling of belonging and oneness among emotionally related individuals, promotes a sense of responsibility and generates a feeling of heroic satisfaction in the genetically non-related donor. Even the donor in Case 3 can be seen to be positively induced in the same way. The recognition of a debt of gratitude puts a person in a special kind of kinship with a person to whom the debt is owed. One can identify a historical event that serves the purpose of binding the donor to the recipient. The event also creates a sense of responsibility and the subsequent donation generates a feeling of heroic satisfaction that proclaims a reciprocal commitment transcending genetic barriers.

The donation of organs under the conditions described in Cases 2 and 3 are more compatible with Confucian conceptions of the family. In Confucian societies, people who are bound by debts of gratitude tend to regard themselves as being related in a way that makes them responsible for returning favors on a continuing basis. The recognition of a debt is recognition of a responsibility for a lifetime. This is the reason why the bonds can be sealed even more tightly than genetic ties. Neither the chauffer in Case 2 nor Danilo in Case 3 can be held back by a rule limiting organ donation to relatives because, within their culture, they consider themselves and the recipients of their donated organs part of the same family.

On the other hand, being committed to individual autonomy, the Western family can easily frown upon the same practices. It tends to see the closeness that is sealed by indebtedness as an example of coercive pressure that gets in the way of free consent. It refuses to recognize the closeness of the relationship and, instead, turns its attention to the eradication of any kind of compulsion.

The main point is that the sense of responsibility that relates to organ donation cannot be confined only to the genetic family. Arising from the realization (1) that there is another person requiring help, (2) that the person requiring help is someone with whom the prospective donor is related in a special way, and (3) that the prospective donor is in a unique position to provide the help needed, this sense of responsibility is capable of flourishing outside as well as inside a genetically based kinship. This conclusion arises more strongly as we are now told that blood relatives are not necessarily more medically suitable donors if only because of the great advances in anti-rejection drugs.

The fact that blood relationship has been rendered minimally significant for the medical success of organ transplants puts into question genetically rooted conceptions of the family. We are only too aware of the complicated nature of human relationships. Genetic relationships are a very special kind, but they do not necessarily take precedence over other kinds or relationships when we are talking about organ donation. Human beings have an immense capacity to relate that goes way beyond blood ties. Hence, eligibility for organ donation should not be made to rest primarily on genetic relationships. While there is a need to be wary of commercially motivated "kinships," this wariness must be tempered by openness to emerging types of human bonds and to flexible conceptions of the family.

Moreover, it is about time that human beings became honest with themselves and looked at the emerging pressures in terms of the challenge to extend the frontiers of human responsibility beyond the boundaries set by genetics and limited reciprocities. The family is not merely a social unit for reproduction but also a social unit for meeting the challenges of disease and infirmity. The age of information technology and biotechnology has imposed new terms for human relationships. We have to recognize the shifting frontiers of our responsibilities in terms of health benefits that "non-relatives" can provide to one another.

The focus on modern reproductive techniques has made us aware of new opportunities to help our own selves. In the field of organ transplantation, we should see new opportunities to help one another as members of an expanding family, whether Confucian or Western. End-stage renal disease is a condition that does not recognize boundaries between Confucian societies and West. While this paper has looked at Confucian and Western responses to developments in transplant technology, it cannot overlook the blurring of differences insofar as actual trends are concerned. As pointed out above, living donors have overtaken cadavers as primary sources of transplant organs in the United States. This is an indication of a growing recognition of an expanding family that is supported by pending legislation giving legitimacy to growing responsibility outside of consanguinity.

On the other hand, Confucian siblings are picking up the notion of individual rights that may be invoked when they feel that they have to resist the coercive expectations of family members. One may see this as a paradox in an age when transplant trends show a move towards responsibility for health concerns that extend way beyond family borders. Perhaps it ought to be taken as a reminder that any effort to balance Confucian societies and Western societies is a complicated exercise.

NOTES

[1] All Kaplan-Meier Patient Survival Rates For Transplants Performed: 1995–2002. Based on Organ Procurement and Transplantation Network data as of March 11, 2005. This work was supported in part by Health Resources and Services Administration contract 231-00-0115. The content is the responsibility of the authors alone and does not necessarily reflect the views or policies of the Department of Health and Human Services, nor does mention of trade names, commercial products, or organizations imply endorsement by the U.S. Government.

[2] Ibid.

[3] Ibid.

[4] Ibid.

[5] Donors Recovered in the U.S. by Donor Type. Based on Organ Procurement and Transplantation Network data, accessed on February 17, 2005. This work was supported in part by Health Resources and Services Administration contract 231-00-0115. The content is the responsibility of the authors alone and does not necessarily reflect the views or policies of the Department of Health and Human Services, nor does mention of trade names, commercial products, or organizations imply endorsement by the U.S. Government.

REFERENCES

Council of Europe (2004a).'International Figures on Organ Donation, Transplantation, Waiting Lists and Family Refusal Year 2003,' *Transplant*, 9–1, 14 [On-line]. Available: http://www.msc.es/profesional/trasplantes/pdf/newstx.pdf. Accessed February 26, 2005.

Council of Europe (2004b), 'International Figures on Organ Donation and Transplantation Year 2003,' *Transplant*, 9–1, 18–20 [On-line]. Available: http://www.msc.es/profesional/trasplantes/pdf/newstx.pdf. Accessed February 26, 2005.

Gjertson, D. & Cecka, J. (2000). 'Living unrelated donor kidney transplantation,' *Kidney International*, 58, 496–497.

Keown, P. (2001). 'Improving the Quality of Life: The New Target for Transplantation,' *Transplantation*, 72, 567–574.

Miranda, B., et al. (2003). "Organ Shortage and the Organization of Organ Allocation," *European Respiratory Monograph*, 26: 62.

Miranda, B. & Matesanz, R. (1998). 'International Issues in Transplantation: Setting the Scene and Flagging the Most Urgent and Controversial Issues,' *Annals of the New York Academy of Sciences*, 862, 129–143.

Organ Procurement and Transplantation Network, 'All Kaplan-Meier Patient Survival Rates For Transplants Performed: 1995–2002 Based on OPTN data as of March 11, 2005' [On-line]. Available: http://www.optn.org/latestData/rptData.asp. Accessed March 15, 2005.

Organ Procurement and Transplantation Network, 'Donors Recovered in the U.S. by Donor Type' [On-line]. Available: http://www.optn.org/latestData/rptData.asp. Accessed February 17, 2005.

CHAPTER 10

THE ETHICS OF HUMAN EMBRYONIC STEM CELL RESEARCH AND THE INTERESTS OF THE FAMILY*

RUIPING FAN

City University of Hong Kong, Hong Kong

I. INTRODUCTION

The research on human embryonic stem cells (ESC) promises to revolutionize medicine in the 21st century. Undifferentiated, pluripotent human stem cells are capable of developing into virtually any body tissue and therefore may be used to replace damaged organ tissues (such as cardiac tissue following a heart attack) or repair currently irreversible injuries (such as spinal cord injuries) so as to recover health. Today's many incurable conditions, such as Parkinson's disease, Alzheimer's disease, multiple sclerosis, and diabetes, may find their reparative therapies in the research on ESC. However, ESC research is remarkably morally controversial in the West. Many people with a Christian background see such research as a grave moral mistake because it has to cause the death of the embryo: in order to conduct such research, scientists have to harvest the stem cells from a human embryo, thereby destroying the embryo.

In fact, some hold a series of moral disagreements against the supporters of this research. First, when the supporters point out that, at the blastocyst stage

---

* The first version of this paper was presented at the "Third International Conference of Bioethics: Ethical, Legal and Social Issues in Human Pluri-Potent stem Cells Experimentations," National Central University, Chungli, Taiwan, June 22–26, 2002. The Governance in Asia Research Centre at City University of Hong Kong provided travel funds for me to attend the conference. In addition, I am grateful to Hon-lam Li for his constructive critiques on an earlier version of the paper. Finally, a revised version was discussed at the "Workshop on Moral and Political Philosophy" organized by the Philosophy Department of the Chinese University of Hong Kong on February 2, 2006. I wish to thank the participants for their very helpful questions, comments and suggestions, especially Hon-lam Li, Norva Y.S. Lo, Joseph Chan, Jonathan Chan, Shih Yuankang, and Ruey-Yuan Wu.

S.C. Lee (ed.), The Family, Medical Decision-Making, and Biotechnology, 127–148.
© 2007 *Springer.*

when the organism is typically disaggregated to create an embryonic stem cell line, the so-called human embryo is only a ball of cells no bigger than the period at the end of an English sentence, opponents argue that, no matter how small it is, a living human embryo is member of the human species: it is already human from conception and therefore has a special moral status. It should, according to opponents, be valued, and not be killed. Secondly, opponents claim that doing such research would desensitize people to the value of human life, threatening vulnerable members of human society. Thirdly, when research supporters argue that there are discarded embryos (from infertility-therapy clinics) available for research, and since they are going to be disposed of anyway the discarded embryos might as well be used for research, opponents contend that such surplus embryos should not have been produced in the first place. Moreover, for the opponents, there is a difference between the embryos' dying and actively killing them. One should not do something that is intrinsically wrong even if good may come of it. Finally, when research supporters attempt to use parental consent to authorize such research, opponents rebut by arguing that parents do not have the moral right to consent to the destruction of the human embryos any more than to the destruction of their own children. To the opponents, killing the embryo in order to harvest its ESC for the benefit of research is morally equivalent to killing a child in order to harvest the child's organs for the benefit of those waiting for organ transplantations.[1]

Although the above arguments of the opponents of ESC have been made with full assurance, it is not apt to offer if one stands outside of the moral context of the Western Christian religion, although Christianity was originally not a Western European religion.[2] In the Confucian tradition, for instance, the full moral significance of an embryo cannot be identified in separation from the context of the family, even if the embryo carries some intrinsic moral value by itself (e.g., simply because it is human, it is morally more important than an animal or a non-human object). That is, in order to decide how to treat an embryo in a specific context, Confucians must consider not only the value of the embryo itself, but also its status in terms of the good of the family. Evidently, Confucians and Christians hold quite different moral perspectives regarding the moral status of a human embryo.[3]

Is it possible to offer substantive answers to such issues as regarding ESC research without reference to the specific moral assumptions of any particular religion or culture, like Christianity or Confucianism? Indeed, this has been the attempt of contemporary Western liberal philosophy and ethics, following the aspirations of the modern Western Enlightenment project. Regarding ESC research, liberal philosophers would attempt to resolve the moral issues by setting up a "pure" rational argument – "pure" in the sense that the argument should be independent of any religious or metaphysical views on the moral status of the embryo. Their general strategy is to offer an account of individual rights so as to place the individual in authority to make relevant moral decisions.

This paper argues that the liberal strategy cannot succeed in offering a persuasive ethical argument for ESC research because it is a version of ethical individualism. The paper contends that Confucian morality is not ethically individualistic, but is a version of ethical familism, which promises to establish a more appropriate moral strategy for addressing ESC research. Towards this end, section II compares how Confucian ethical familism differs from liberal ethical individualism in moral exploration. Section III demonstrates how liberal ethical individualism is a one-sided ethical strategy that cannot lead to proper ethical conclusions regarding ESC research. Section IV explains how Confucian ethical familism offers a two-dimensioned ethical strategy and at the same time does not engage utilitarian maximization. Section V lays out a Confucian moral casuistry regarding ESC research and explores how particular Confucian views and conclusions can be justified through its virtue-based, two-dimensioned moral approach. Section VI contains concluding remarks on different moral traditions.

II. ETHICAL INDIVIDUALISM VS. ETHICAL FAMILISM

Confucian morality is a type of ethical communitarianism, not ethical individualism. To recast this morality for analysis and comparison, let me summarize it into the two following principles:
(1) Both individuals and their communities ultimately count;
(2) Individuals do not count equally.
Before elaborating on them, it is necessary to have a word about the debate between individualism and holism in the contemporary philosophy of social science. Methodological individualism holds that (1) the whole is nothing but a collection of individuals related to one another in a certain way (the ontological thesis), (2) social concepts are definable in terms of the concepts that refer to individuals and their relations (the conceptual thesis), and (3) adequate social explanations must refer solely to individuals, their relations, dispositions, etc. (the explanatory thesis) (see, e.g., Little, 1991, Chs. 9, 11). Methodological holism may well accept the ontological thesis, yet rejecting the conceptual and explanatory theses. Indeed, the debate in the methodological area has been intense. On the other hand, however, the debate in the ethical area has almost been silent. This is probably due to the decline of the Platonic and Aristotelian holistic ethical views in the modern West that has given a dominant position to ethical individualism.[4] Accordingly, in order to explicate the Confucian ethical principles clearly, it is helpful first to compare them with liberal individualist ethical principles. Indeed, at the core of liberal ethics is ethical individualism, whose major principles can be summarized as follows:
(1)' Only individuals ultimately count;
(2)' Individuals count equally.[5]
The liberal principle (1)' means only individuals have intrinsic values – values that count by themselves, without relying on any other things. This principle entails the instrumentalist view of community: a community does not have any intrinsic value. The worth of a family, an organization, or a state, for example, depends on its contribution to the individuals out of which it is constituted. That is, a community is valuable insofar as it serves the values of individuals. Accordingly, under ethical individualism, the interests of community are considered in terms of the interests of

individuals, while the interests of individuals cannot be considered in terms of the interests of community.[6] From liberal contractarian theories, individuals are ends, while communities are means: means should be constructed, revised, or rejected according to ends.

The principle (2)' indicates a liberal egalitarian position: individuals should be treated as equals. Although this position does not necessarily support an equal distribution of income, it discloses a basic liberal idea of equality: the interests of each member of a community matter equally for the community. In other words, it requires that each member be entitled to equal concern and respect and that each member's interests be given equal consideration. In practice, this egalitarian thesis usually leads to some specific versions of concerns for equal treatment, such as equality of opportunity, equality of resources, equality of capacity, or even equality of welfare, although liberals disagree on which is the more appropriate version of equality. It also entails and supports a series of equal individual rights, which can conflict with each other in practical contexts. Regarding the moral issues relevant to ESC research, there are two often referred but mutually contradictory rights under the liberal principle: a right not to be killed and a right to control the use of one's own body.

Confucian communitarian ethics as summarized in the beginning of this section sharply contrasts with those liberal individualist theses. First, although Confucianism grants an independent intrinsic value to the human individual (to wit, a human being is highly valuable simply because he/she is a human being per se, not because he/she has higher capacity than other animals or is useful to something else),[7] yet it holds an anti-instrumentalist position of community. Confucians understand that individuals live in different types of communities, including geographically located communities (such as families, villages, cities, states) and non-geographically located communities (such as associations and religions). Those communities overlap with each other in terms of their constituting members, and each of them carries different moral significance for different people. Moreover, people naturally see different types of communities as morally primary for individuals – primary in the sense that the moral claims and interests of these communities trump those of other communities. For instance, Japanese people see their nation as their primary community; Christian believers see their church as their primary community. For Confucians, the family is the primary community for every individual. In this sense, Confucians can be termed ethical familists.[8]

Confucians have to disagree with liberal individualists regarding the ethical value of the family. Liberal individualists see the interests of the family as nothing but the sum total of the interests of individual members. Even if they like to count all members – including not only current existing members but also members living in the past and members coming in the future, the moral focus is evidently on currently existing members. In contrast, Confucian familists would see the interests of the family as greater than the sum total of the interests of currently existent individual members because they must include the well-being of deceased ancestors and future descendents. This is why the rituals for memorizing the ancestors and practicing the children's virtue of filial piety have been so foundationally important in the Confucian tradition. As Confucians see it, the family life embodies the right and flourishing way of human existence. The family substantializes the basic human relations, which are irreplaceable for individuals to pursue human flourishing. In contrast to a contractarian understanding of the family, Confucians see the family to be in nature non-voluntary for the individual to begin with: one is naturally born into the parent-child relation as well as other family relations without giving voluntary consent in the first place. The best one can expect is to strive for a good life together with these naturally already attached relatives in harmonious relations. However, when Confucians hold that both individuals and families are intrinsic values or ends, it is incorrect to say that individuals are merely means for families, just as it is incorrect to say that families are merely means for individuals. An analogy may well be drawn with the case of Aristotle on the relation between the citizen and the polis: although the polis is a natural end, there is no antithesis between the citizen and the polis, for the polis is ultimately meant for the fulfillment and perfection of the citizen. In the Confucian case, both individual and family ends should be integrated into a coherent system grounded in a doctrine of the virtues.

This Confucian emphasis on the family, however, may fail to distinguish Confucian ethical familism from liberal ethical individualism at a deep level. This emphasis indicates that, given human nature as it is, the family is naturally necessary and valuable for the individual to perfect him/herself. Thus, it appreciates the family ultimately in terms of individual flourishing. However, the family in the Confucian tradition carries a special moral value or dignity independent from the value or dignity of the individual: the existence of the family reflects a profound moral structure and significance set by Heaven (tian). This Confucian thesis has important normative moral implications. Negatively, it implies that we should not create such human individuals who would not require the family for moral flourishing, even if this has become technologically possible through genetic engineering. That is, even if we can technologically create "new" human beings who would possess a dramatically different physical and psychological make-up from ours so that they would no longer need to fulfill their "natural" ends through the family and therefore the family would "naturally" disappear, we should not do so because the family, as Confucianism holds, carries intrinsic moral values independent of individual values. It is morally bad to lose familist intrinsic values. Positively, the Confucian thesis implies that we should attempt to maintain, strengthen and even enhance the traits or emotions that are suitable for the family life. In short, genetic engineering makes it possible to shape individuals in future. But what kind of individuals should be shaped depends on what values are taken to be ultimately fundamental. Unlike Aristotlelians who may not care about the family seriously, Confucians want to shape those individuals who carry the appropriate sexes, emotions and intelligence for leading the lives of the family.

In short, Confucianism holds that the family carries intrinsic goods, which are irreducible to the goods of the individual, no matter how individual goods are defined.[9] I would, for the purpose of comparison with liberal individualist ethics, reconstruct a Confucian perspective regarding where individual and family goods reside, without offering systematic Confucian textual evidence (which could be left for another paper). Confucians would see the good of the individual lie in one's survival and health, a complete family life (including being loved by the parents when young and being cared by the children when old), as well as the satisfaction of one's main, reasonable desires or preferences. Obviously, these three different types of components can conflict with one another in particular contexts. Suppose, in every context, the interests of the individual can be worked out by balance. Then, importantly, individual interests could conflict with family interests in circumstances. I would summarize the interests of the family in the Confucian tradition as the integrity, continuity and prosperity of the family. The integrity of the family consists of the virtuous activities of individual members, their shared decision-making concerning important issues, as well as the good reputation of the family. Without these elements the family would be morally incomplete. No doubt, the Confucian family is heterosexual and patriarchal. The continuity of the family consists in that the family must have a son to succeed the family. This is why the most unfilial thing for Confucians is having no posterity (e.g., *Mencius* 4A.16). Finally, the prosperity of the family lies in both material wealth and harmonious relations (*he*) in a family. In addition to achieving affluent resources for the members' welfare, accomplishing appropriate and pleasant family relations (*ren lun*) through everyone's virtue cultivation is vitally important for the Confucian life to flourish. The process of their harmonious shared-determination is precisely a process of communication, reciprocation, compromise, and voluntary sacrifice. It is first and foremost the shared experience of the common family life that distinguishes out the Confucian way. It is highly contextual, poetic, and holistic. Its uniqueness, mystery as well as the holistic family values involved require comprehensive narratives to comprehend and cannot be fully approached through speculative discourses.

However, when individual interests come into conflict with family interests, there is no simple Confucian formula that requires the sacrifice of individual interests for the family interests, as some may have conceived. Indeed, there is no such clear-cut formula as to whether individual interests should submit to family interests, or vice versa. No doubt, like Aristotelian ethics, Confucian ethics can be taken as an educational system: individuals should be nurtured to harmonize their individual interests with their family interests. In this sense, perfectly realized individual goods would not stand in conflict with the family goods. But individual goods are often not perfectly realizable in reality, and so are family goods. It may well be the case that, all things considered, an individual would be better off if he would not get married and would not have children, but this would not be good for the family. Then what should be done? The best that the Confucian families can pursue is to explore what a virtuous decision and activity would be in such a case.[10] One has to turn to specific Confucian moral elaboration and casuistry in order to understand this Confucian stance in its full sense (see Sections IV and V).

The Confucian anti-egalitarian view of individuals holds that individuals should not be treated as equals; rather, they should be treated as relatives. The five basic human relations admired by the Confucian tradition are not only meant for acquaintances, but are for all people (including strangers) to form – strangers, once met, should be placed into one of these five relations. The ruler-ruled relation is like the parent-child relation. When people are good friends, they come into the relations of brothers or sisters (namely, older and younger). For Confucians, no one is unrelated. The difference is only that some are close relatives, while others are remote relatives (Chan, 1963, p. 70). Some might want to argue that an "egalitarian" level can still be teased out from this Confucian moral account. The requirement of treating people as relatives, they may contend, must include an egalitarian "threshold" below which people are no longer treated as relatives. However, a requirement of equality is never an emphasized point in the Confucian familism. First, relatives are in nature unequal. What is important for Confucians is not to emphasize that the father and the son should treat each other with equal rights, but that the father should treat the son with the virtue of kindness (ci) and the son treat the father with the virtue of filial piety (xiao) – that is, they should cultivate different, *unequal* specific virtues. Second, the closer the relation one has with another, the more consideration one should give to his/her interests. It is morally wrong for Confucians to believe that one should equally consider the interests of one's son and another's son. Rather, Confucianism holds that one ought to consider the interests of one's son more than another's son. Third, as each individual exists in distinct, specific situations, it does not make sense, for instance, for a Confucian mother to say that she should treat a fetus in her womb and a child already born as two equals. What is proper for her is to treat them in different manners suitable to their respective characters and contexts as well as their relations with others in terms of specific Confucian virtues.

What about politics and policy? Shouldn't government treat all citizens as equals and consider their interests equally? Again this is missing the real point in the Confucian concern. The Confucian principle of treating people as relatives upholds a harmonious (*he*), rather than an equal, political system. First, the concept of citizenry is already a narrow concept, insufficient for taking care of the Confucian ideal of all-under-heaven (*tian xia*). Under the Confucian harmonious system of the state, the young should be cared for, the elders should be respected, and the foreign should be attracted to join. Their interests should always be considered in different ways. Even for adult citizens, "to be treated as equals" is still not a good idea, because individuals are not equal in learning and in practicing their virtues. As Mencius sees it, although individuals have equal moral potential (that is, everyone carries the seeds of the virtues) to develop themselves, they accomplish this differently through unequal efforts. Confucianism is a type of ethical and political elitism in this regard: individuals should be treated according to the virtues they have nurtured and achieved – the more virtuous the person, the more respect he/she should receive in society.

In short, the Confucian moral principle that the family has intrinsic moral value opens up a new moral dimension for ethical exploration – the balance of individual and family interests. Because the family has intrinsic value, each family member should also take care of the family interests in dealing with their individual affairs. Since some individual affairs significantly affect family interests, they become matters of interest for all family members and are open to common exploration in the family. This virtually leads to the moral model of family shared-determination (rather than individual determination) in the Confucian tradition. All important individual issues, such as education, marriage, and health care, should be decided by the whole family for each family member so that a decision will not only be good for the individual, but also good for the family. When family interests are at stake, it is inappropriate for one to declare that "this is my own business" or "please leave me alone." The value of shared-determination is implicit in the normal Confucian family life. Moreover, since the individual also ultimately counts, the family must seriously take into account an individual's view when making a decision for him/her. In this regard, shared-determination integrates self-determination and moves it to a higher level of moral decision-making. In contrast, the liberal individualist model of individual interests as well as self-determination is one-sided. It is no surprise that liberal individualist ethics cannot offer a persuasive argument regarding ESC research, to which we turn now.

III. ONE-SIDED MORALITY ON ESC RESEARCH

Liberal individualist ethics is the product of the modern Western Enlightenment movement. Its basic principles are already implicit in Kant's famous statement: individuals should always be treated as ends, and should never be treated merely as means. Kant failed to recognize that some communities, like the family, are essential for individuals to pursue human flourishing. Worse yet, while the Kantian notion of individual autonomy carries the meaning of universal legislation through his formulation of rationality as universalizability, contemporary liberals, facing the ever-increasing diversity and plurality of moral values (namely, there are incompatible and incommensurable moral visions and conceptions of the good life competing with each other in contemporary society), have reinterpreted the Kantian autonomy in terms of individual liberty or self-determination: every individual is in authority to order his/her life and decides his/her acts as he/she sees fit, as long as it does not harm others (the self-determination thesis). Hence, the classical liberal autonomy has become contemporary liberal self-determination, and one-sided, self-regarding morality. This morality is hard to handle genetic ethical issues because these issues are inevitably family relevant and other-regarding.

Specifically, the self-determination thesis has two versions: weak and strong. The weak version holds that individual self-determination should not be coercively interfered with by others or society, although individual self-determination is not an intrinsic value by itself. That is, under this version, the individual is solely in authority to make decisions concerning his/her action, and society is limited from intervening with this decision, but society does not have any obligation to encourage or promote individual self-determination. On the other hand, the strong version of the thesis holds that exercising self-determination is an intrinsic value for the individual. Respecting individuals requires not only not interfering with their self-determination, but also positively strengthening their capacity for exercising self-determination. Hence, the weak and strong versions differ regarding whether self-determination is an intrinsic value to be promoted.[11]

Accordingly, liberals would find it misleading to emphasize the interests of community (such as the family). For them, community matters only in terms of how it affects individual interests, and the form or structure of community should be contractually revised according to individual values, wishes, and agreements. Thus, contractual relationships become the primary, normative social relationships, while traditional relationships based on affective networks are understood as the epitome of pre-modern backward forms of social relations. Liberals focus not on establishing and pursuing a notion of the common good for a community; rather, they regard the common good in terms of the good of most of the community's individuals or as that good for individuals, which can only be realized through cooperation with other individuals. All of this can be understood contractually. To support such contracts, liberals need only set up principles of justice to ensure individual liberty and rights so that individuals can freely undertake their life plans. Indeed, this has been the major task of liberal philosophy in the contemporary world.

How does this liberal moral and political understanding bear on ESC research? Obviously, the answer will depend on how liberal ethics takes the status of human embryos. When liberals accent individual liberty or self-determination, typically they have in mind adult human individuals – only these are human agents capable of practicing self-determination and making contracts with each other. For contemporary liberals, respecting the life plans of such agents seems to be the only way of treating them as ends. Indeed, this individualist contractarian model seems to work well with adult individuals in a pluralistic society. The maxim is: do not put your nose into my business if it does not involve you, and if it involves you, I shall seek your consent in the first place. However, a problem occurs with the issue of reproduction. When one and one's spouse decide to produce a human embryo, they are deciding to bring another individual into being. They cannot seek consent from the individual beforehand because it has not existed yet.

Does this mean that adult individuals are morally at liberty to reproduce an embryo for whatever purpose they wish (such as to do research on it or simply to kill it later for fun) or in whatever way they prefer (such as through the traditional way of sexual intercourse or through contemporary technical procedures like in vitro fertilization)? May they create a baby just as they may make a desk? Indeed, one may make a desk for whatever purpose one holds (such as using it as a reading table, a computer shelf, or simply destroying it later for fun) and in whatever way one prefers (such as traditional manual way or some modern technical method). Some would quickly add that producing an embryo should differ from producing a desk because an embryo, not a desk, possesses the potential of developing into an adult human individual that has the capacity of self-determination. The question is:

does the recognition of this potential set any moral constraints on adult individuals' purposes for, or ways of, producing and treating an embryo in a liberal account?

From my view, the liberal individualist answer to this question will have to be "no", whether one holds the weak or strong version of self-determination. If one holds the weak version, one wants to insist that one's decision and action, whatever they are, should not be coercively interfered with by others, as long as they do not involve others. If my partner and I cooperate to produce an embryo for research, then only my partner and I are in authority to make such a decision and action, because the only other one involved here is the embryo that does not have any capacity of self-determination. Although it has the potential for this capacity, this potential cannot bind my partner and me, because the weak version of liberalism does not take that capacity as a value, much less the potential. It is up to me and my partner – up to the particular personal, moral or religious view that we happen to hold regarding the embryo – to decide what I want to do about an embryo. If I think an early embryo created for ESC research is nothing morally special than a desk designed for a carpenter's use or experiment, no one is in authority to stop me from doing it. This is to say, under the weak version, the recognition of the potential of an embryo for self-determination should not set any real moral constraint on individual purposes for, or ways of, producing an embryo. Accordingly, ESC research (no matter either the embryos involved are from in vitro fertilization procedures or from on-purpose embryonic clones) should be ethically permissible in society.

On the other hand, if one holds the strong version of liberalism, it seems that a different structure of moral deliberation emerges. Here self-determination or individual autonomy becomes an intrinsic value to uphold. Of course, it is not that self-determination or individual autonomy must be the only intrinsic value held by liberals. Liberals may hold other intrinsic values, such as friendship, happiness, knowledge, and even some familial and religious values, depending on particular individual liberals. However, liberals cannot be perfectionists. Even if they hold a set of intrinsic values in addition to individual autonomy, they must stand ready to trump those values by individual autonomy if they are in conflict. This is why liberals, but not the libertarians holding the weak version, have to take that voluntary slavery is morally mistaken. That is, liberals have to hold self-termination as a dominant intrinsic value (Kymlicka, 2002). The liberal reasoning regarding the appropriate way of raising children constitutes an illustrative example in this regard: the primary purpose of liberal education is the promotion of children's capacity for self-determination.[12] In this regard, the recognition of the potential of an embryo for self-determination should set a moral constraint on individual purposes for, or ways of, producing and treating the embryo.

However, the strong version of liberal argument on the value of the embryo cannot logically lead to this conclusion. First, it is clear that self-determination is a dominant intrinsic value (that is, in conflict with other values, self-determination dominates). Moreover, it is also clear that the life of an individual that exercises self-determination is dominantly intrinsically valuable (that is, without an overriding or compelling reason, the life should not be taken). This corollary is derivable because the life of such individual is both the necessary and sufficient condition for self-determination. Then what about the life of an embryo that has the potential of developing into the life of an individual that can exercise self-determination? Would it only be an instrumental value because it is only a necessary condition for the life of the adult individual? Perhaps the liberal could distinguish internal and external necessary conditions: the life of an embryo is an internal necessary condition, while other things, such as temperature, water, nutrition, etc., are only external necessary conditions, for the life of the adult individual, since it is the embryo, not any other thing, that can develop into the adult individual in a suitable environment. In this sense, liberals can take the embryo as intrinsically valuable.[13] Nevertheless, they cannot take it as dominantly intrinsically valuable – as one intrinsic value, it has to be balanced against other intrinsic values through relevant adult individuals' practice of their self-determination. In this way the strong version liberal view has to collapse into the weak version liberal view: eventually, it is up to the actual desires, wishes and preferences that adult individuals happen to hold to decide how they would balance the intrinsic values so as to decide what should be done with an embryo, such as whether it should be created and used for ESC research.

Consequently, in response to the question of whether recognizing the potential of an embryo to become a full individual with the capacity of self-determination sets any moral constraints on our purposes for, or ways of, producing it, the ultimate answer from either the strong or the weak versions of the liberal self-determination would be "no." This would be the case, if the above argument is sound, no matter whether particular liberals recognize the full sense of their theses.

From a Confucian's view, the argument from the weak version seems incomplete: on the one hand it grants individual self-determination so high a status that society may not interfere with it at all, and on the other hand, it does not take it as a value for promotion. Then it is hard to understand why self-determination should possess such an absolute position. On the other hand, the argument from the strong version seems incoherent. It assigns a dominant intrinsic value to the individual of exercising self-determination, but it ends up without setting any serious moral restriction on the production or use of an embryo that carries the internal potential of becoming the individual of exercising self-determination. Both liberal answers are morally problematic. If their arguments are taken, it is not only that society should not prohibit ESC research, but neither should society place a brake on late-stage abortion or even infanticide, because all the individuals involved here do not have the capacity for self-determination. This is ironic in that liberal ethical individualism holds only individuals to have intrinsic values. For Confucians, this circumstance indicates that liberal ethical individualism does not have sufficient moral resources to handle the moral issues such as ESC research.

IV. THE CONFUCIAN TWO-DIMENSIONED MORAL STRATEGY:
NOT UTILITY MAXIMIZATION, BUT VIRTUE PURSUIT

Confucian familism offers rich moral resources for considering reproduction and ESC research. On the one hand, reproduction is of individual interest: everyone has specific expectations, wishes, and preferences regarding one's reproduction. On the other hand, one's own expectations should be mediated by the interests of the family. Since issues relating to reproduction significantly affect the interests of the family, Confucianism holds that every family member must participate in the process of shared-determination of the family. Family members together explore the specific implications of individual acts for the integrity, continuity and prosperity of the family as well as for the interests of the individual at stake. Consequently, when liberal individualists solely rely on individual values to decide the manners of their reproduction, Confucian familists appeal also to the family values, thus the Confucian two-dimensioned moral strategy. Methodologically, such a two-dimensioned morality assists to overcome the extreme conclusions drawn by the liberal one-sided moral views as shown in the last section. If only individuals ultimately count, individuals must confront individuals regarding their interests in the context of reproduction or ESC research. When their interests conflict, a gridlock is formed and cannot be resolved unless through an all-or-nothing strategy: either some individuals do not count and therefore anything can be done to them, or every individual counts and therefore, nothing can be done. The Confucian introduction of both individual and family values to moral elaboration offers the opportunity of appealing to familist interests to tip the scale at an appropriate point.

But Confucian familism is not a type of utilitarianism aiming at the maximization of family interests in individual reproductive acts. First, Confucian familism does not hold ethical individualism as utilitarianism does. Although utilitarianism is teleo-logical and liberalism deontological (so that they are usually distinguished as two quite different forms of ethical theories), utilitarianism shares ethical individualism with liberalism: only individuals ultimately count. Second, Confucianism does not hold a value reductionism as utilitarianism does. Even if utilitarians do not have to strictly adopt the interest-maximization formula as a basic moral standard (that is, utilitarians may not have to conclude that, when individual interests conflict, those individuals with less interests should be sacrificed for those individuals with greater interests in order to maximize individual interests), they have to appeal to the quality-of-life view to locate the value of human life in such qualities as consciousness and rationality. In this reductivist way they can conclude that (1) some lives are of greater worth than others, and (2) some lives are not worth living, and thus it is in their "best interest" to die.[14] In contrast, Confucians do not take the value of the individual to be reducible to a set of qualities. For Confucians, a human life has a sacred property because it has been invested with noble moral seeds by Heaven, as Mencius explicates. Thus, its value cannot be fully reduced to empirically identifiable and comparable traits.

How then would Confucians resolve the possible conflicts between individual interests and family interests? For instance, if one is the only son of a family, the best interests of the family, all things considered, are for him to get married and reproduce children, while his own best interests, all things considered, may be remaining single and having no child. In many cases, the best interest of the embryo is to be brought to birth, but the best interest of the family, all things considered, is to terminate the pregnancy. To address such conflicts, Confucianism would not claim that family interests always transcend individual interests, or vice versa, as mentioned in Section II. Confucians must appeal to virtue to guide their actions. In pursuing any interests, either individual or familial, the fundamental Confucian moral requirement is that one's action ought to be appropriate, fair, or righteous (yi). Indeed, that is the basic teaching of yi – the Confucian virtue of justice (*Analects* 4.10, 17.23; *Doctrine of the Mean* 20.5). Yi is the character that enables one to concentrate on the right or the righteous in facing benefits, especially treating others appropriately (*Analects* 4.16, 16.10, 19.1). Accordingly, such cases as killing one person in order to use his organs to save five other persons may well be posed to challenge the moral property of utilitarianism – if an all-things-considered "impartial" calculation shows that such killing maximizes interests, utilitarianism should conclude that it is morally right to do so. But they cannot be used to attack the Confucian morality because such killing violates the basic Confucian virtue of righteousness (yi). Crucially, Confucian familists would calculate the family interests only after the circumstances have been judged not violating the virtue of yi.

This is to say, it is a misinterpretation that Confucianism would certainly support the sacrifice of the embryos in ESC research because such research promises huge likely benefits to families, given that Confucianism takes family interests as intrinsic values. In ordinary moral lives, Confucians must follow the established rituals (li) to make authentic behaviors. In moral crises (such as in a grave conflict of interests between individual and individual or between individual and family), the situation must be weighed (quan) through moral discretion according to the virtues (*Analects* 9.30; *Mencius* 4A.17). Such moral discretion is primarily not utilitarian. How should the consideration of family interests play a role in this discretion is a highly context-dependent issue which requires a Confucian moral casuistry.

V. A CONFUCIAN MORAL CASUISTRY ON ESC RESEARCH

Classical Confucian literatures never directly address such issues as abortion or others closely relevant to ESC research. It seems that classical Confucians were generally silent on the moral status of human embryonic life. Here I am not in a position to give a full account to this issue. But two observations may help. First, Confucian metaphysics is not such that the moral essence and status of personhood can be found in a particular act of divine creation. Instead, Confucians see the biological mechanism of the human genesis, growth and birth through the parents' acts as itself the manifestation of the good mandate of Heaven (tian). That is, Confucians find that the way (dao) of Heaven has been fully presented by the parents' proper reproductive conduct, without any additional divine infusion required. "It is man who can make the way great; it is not the way that can make man great" (*Analects*15: 29). "Unless there is perfect virtue, the perfect way cannot be materialized" (*The Doctrine of the Mean* 27.5).[15] Accordingly, Confucian society has relied on the parents' good intentions and appropriate actions to complete a recreation in accordance with virtue. It is impossible for Confucianism to have an explicit prohibition on abortion from a personal God as in the case of Judeo-Christianity. Moreover, it is not that Confucianism does not have capacity to appreciate the moral significance of early human embryonic life. It is rather that the Confucian two-dimensioned morality must bring every individual life into the good of the family. It is no doubt bad to conduct abortion or sacrifice an embryo in another manner, because the embryo, as a human life, has intrinsic value in the Confucian tradition. But in a conflict of interests between an embryo and an existing family member or between an embryo and the family as a whole, it is neither morally proper nor practically useful to stress an absolutely inviolable individual worth of the embryo. In moral discretion, Confucians must refer to an embryo's standing in terms of the family good. This might be why they have not found focusing on an independent moral status of embryonic life to be morally sensible.

Confucian concerns with family values can promote individual values. For instance, Confucianism would take it morally wrong for a deaf couple to attempt to secure a deaf (rather than normal) child through artificial reproductive procedures, because such parental attempt damages the family interests. It is difficult for ethical individualists to argue that such attempt harms the child's individual interests, because the child would not have been existed if the parents were not allowed to bring about a deaf child via reproductive technologies. So the harm inflicted would become a harm to nobody within the individualist ethical deliberation. In contrast, in the Confucian two-dimensioned moral structure, the harm is definitely a harm to the integrity and prosperity of the family at stake. Intentionally making one's child deaf is destroying the family integrity because such conduct is unrighteous (bu-yi) to a child, whoever this child will be. It is impairing the family prosperity because it is difficult for a deaf and dumb child to gain wealth for the family. Indeed, Confucians would find that such a case heuristically indicates the moral impoverishment of the rights-based individualist ethics.

Nevertheless, it is not right to insist that a human embryonic life should never be sacrificed for the family interests. Rape cases are important examples. Those ethical individualists who hold the sanctity-of-life view have intractable difficulties in arguing for abortion in the rape case because the embryo is an innocent life, plus that the mother's life is not threatened by the presence of the embryo in this case (so that argument based on self-defense cannot work well). However, while Confucians recognize that the rape case is entirely evil and tragic and that abortion is generally bad, they can morally support the abortion in the rape case because bringing the embryo to maturity would unrighteously generate a terrible strike to the good of the family. As stated in Section II, part of the family integrity consists in the good reputation of the family. If the family decides to bring the embryonic life to maturity in the rape case, it would be like perfecting the aggression of that life imposed by the rapist into the family. This would impair the family's reputation. Moreover, this would damage the continuity of the family not only because it injuries the blood-purity of the family, but it destroys the sovereignty and mastery of the family on the issue of reproduction through the practice of shared family-determination. Indeed, although biological descendents are favored in the Confucian tradition, adoption through a harmonious family shared determination has always been morally acceptable. The evil of rape will have destroyed all this mastery and harmony of the family. Finally, the individual, if brought up, would be caught in a relational disaster. It would be very difficult for the individual to melt into the family and contribute to its prosperity. Accordingly, it is morally right for a Confucian family to abort the embryo in the rape case.

Indeed, the Confucian consideration of the family interests can help them handle certain difficult cases regarding reproduction in a virtuous, non-utilitarian manner. For instance, if either the fetus or the mother, but not both, can be saved, and if in order to save one, the other must be directly taken, then Confucians would hold that the fetus should be taken in order to save the mother. This is because, for Confucians, even if the fetus and mother carry equal individual values, the mother is more important in terms of the family values. Evidently, those ethical individualists holding the sanctity-of-life view have an enormous trouble in handling such cases. Some of them have argued that since both the mother and the fetus are innocent individuals carrying equal values, a fair random method should be taken to decide who should be saved in such a case, if the utilitarian maximization is not accepted as a fair strategy (see, e.g., Brody, 1972, p. 340). Confucians would find this "random" method morally stubborn, unwise, and unrighteous (bu-yi): it is the mother, not the fetus, who has contributed to the family values so that it is only righteous (yi) to save her life, even if her life is of equal individual value as the embryo's life. This primarily is not the utilitarian maximization, because the first important idea is not to calculate the total amounts of the values at stake. It might be the case that the fetus, if brought up, would contribute to the family more than the mother had. Indeed, the fetus might turn out to be the only child that the family could have. Still, these considerations cannot be legitimately integrated into the Confucian moral deliberation of the family interests according to the virtue of righteousness (yi), because righteousness requires, among other things, treating everyone appropriately in facing benefits. It would be unfair not to save the mother's life in order to gain more family interests in such a case, because appropriate reciprocity is the essential Confucian sense of treating everyone appropriately. Indeed, it would be morally disloyal and shameful not to choose to save the mother's life in this case because, for Confucians, this is equivalent to not appreciating or reciprocating the well-intended contribution that the mother had already made to the family. Indeed, Confucianism would take such appropriate reciprocity so important that even if the scenario is such that the fetus will survive and the mother will die if nothing is done, it would still be justified for Confucians to take the fetus in order to save the mother.

This analysis of the rape case can be used as an analogical, paradigmatic case for therapeutic cloning: if nothing is done, a current family member will be lost due to an incurable disease; given her contribution and importance to the family, it is justifiable for the family to create an embryo by using her own somatic cells to offer embryonic stem cells so as to save her life (suppose this has been technologically possible).[16] Some may want to argue that the two cases are morally different: in the first case the fetus is not created for sacrifice, but in the second case that is exactly the purpose – so therapeutic cloning involves not only directly taking a human embryonic life, it also involves creating it only to destroy it. Therefore, for such objectors, therapeutic cloning is morally wrong even if directly taking an embryonic life can be justified in the first case. However, I believe Confucians could successfully defend therapeutic cloning by appealing to a moral insight that has a family resemblance with the Roman Catholic bioethical doctrine of double effect: in certain circumstances, one should intend to do one thing while fully recognizing that a second effect will also result. In the case of therapeutic cloning, the family intends to rescue a family member by setting up a process of cloning and stem cell harvesting while nevertheless foreseeing that the embryonic life will be taken. Like the Roman Catholic account of double effect, the second effect, in this case, the destruction of the embryo, has to be considered an evil. But the entire action, unlike in the Roman Catholic account of the issue, can be justified in the Confucian account: it does not violate the Confucian virtue of righteousness because it is well intended and resulted to save a family member's life.

Can these thoughts be drawn on to justify ESC research? It is one thing to take an embryonic life in order to save a person; it is quite another to take it in order to do research. On the other hand, promising stem cell therapies would never be accomplished if sufficient research will not be conducted. Fortunately, there are different types of ESC research. I want to argue that Confucians would have no problem with the type of ESC research, which uses the extra embryos left over from in vitro fertilization procedures, as long as the latter are morally justified. Indeed, in vitro fertilization itself involves sophisticated moral problems. This essay assumes that Confucians can accept in vitro fertilizations, as long as there is no infliction on the family values, especially the integrity and continuity of the family. This requires that a zygote formed through technological aid should not come from anyone other than the husband and the wife. This also implies that surrogate motherhood is morally problematic.[17] Having met these requirements, the in vitro fertilization technology can be considered by Confucians as an acceptable extension of human physiological capabilities for reproduction, although it involves the disassociation of unitive and reproductive sexuality between husband and wife. In this case, to make surplus zygotes in order to pursue a better success rate of reproduction does not need to be taken as violating the virtue of righteousness (yi). A double-effect argument of the similar structure as I offered above can also be adopted for this case. Here the primary intention of the family is to secure a child for the family, and this intention is morally admirable in the Confucian tradition. After the reproduction is successful, it should be up to the parents to determine that the extra, "frozen" embryos be discarded rather than continuously implanted for developing more children. In any case, in the ordinary course of natural reproduction, a great number of zygotes never implant, and the family should have moral authority to decide how many children to have.

If it is morally acceptable to discard the extra embryos left over from in vitro fertilization, it should be morally permissible to use them for ESC research without violating the Confucian virtue of righteousness. Some may want to argue that it is one thing to let these embryos die by abandoning them, but quite another to kill them by harvesting their stem cells for research: the probity of letting an embryo die cannot automatically justify the probity of direct killing. However, the following considerations should be sufficient to justify the research: first, although these embryos are used in ESC research, there is nothing for them to suffer or lose – they do not suffer in such research, and they are going to die anyway in a relatively short time. Second, the intention to conduct the research is good. Finally, there is everything to gain from such research – it promises great possible benefits to individuals and families in future. Accordingly, the family should have moral authority to decide such extra embryos to be used in ESC research.[18]

Finally, what about the type of ESC research that uses the human embryos – either clones or non-clones – who are created for the purpose of research? I think it is enormously difficult for Confucian ethics to argue that this type of research does not violate the Confucian virtue of righteousness. This case differs from the case of therapeutic cloning in a significant respect: there the intention is to prevent the loss of an existing family member (so that there is an explicit urgent family interest at stake); here the intention is to develop research in order to benefit some individuals and families in the future (so that there is not an explicit, urgent family interest at stake). Could one have a strong Confucian reason to donate one's sperms, ova or somatic cells to create embryos for research? Could the killing of an embryo in this case be justified based on the Confucian familist concerns? First, the intention to do research that benefits some individuals and families in the future is not so compelling a reason for a Confucian as the intention to rescue a particular family member calling for the treatment at this moment. If you were a Confucian, it would be at least morally strained or even disingenuous to claim that to create an embryonic life from your family and sacrifice it for possible benefits to individuals and families in the future is precisely righteous (yi). Moreover, the possible gains achievable from this research is much less certain than the gain that could be obtained from therapeutic cloning when it is already technologically reliable. In the therapeutic case, you would almost be certain that a current family member's life would be saved, while in the research case the embryonic lives used in the research might all be sacrificed in vain. All these considerations indicate that it would be very difficult to construct a convincing double-effect argument from the Confucian moral resources to support the killing of embryos in this research case.[19]

VI. CONCLUDING REMARKS: LIBERALS AND ARISTOTLE, ABRAHAM AND CONFUCIUS

Liberal individualist morality is too general to offer rich guidance on moral issues such as ESC research. It relies on individual self-determination to deal with such issues, without being embedded in any full-fledged moral perspective or a complete conception of the good life. But it is next to be incoherent. On the one hand, it insists that only individuals have ultimate values, restricting community to an instrumental function. On the other hand, however, it has to come to a position that is almost impossible to put a brake on the destruction of human embryonic individuals in ESC research. This indicates that liberal ethical individualism is intellectually and morally infertile, in addition to the irony and sophistication of human moral lives.

A return to the Aristotelian polis is not a right solution. While the polis is a natural necessity for individual perfection, its ultimate values depend on the fulfillment of human nature as political animals. What about changing human nature? In particular, what if future genetic engineering enables humans to reproduce themselves not through the union of two sexes but to have humans who do not live in the institution of the family in order fully to perform their functions as citizens (that is, the Platonic totalitarian society could gain a biological support from possible new technologies)? Shouldn't Aristotelians support such change? But Confucians cannot. Confucianism recognizes that the family as the primary community of individuals carries ultimate intrinsic values, which should never be changed. The profound love embodied in the structure of the family is so foundational, valuable, and irreplaceable that the moral nature of the family would never be tradable. Eventually, the Aristotelian friendship is only a one-sided, individualistic self-love, but the Confucian humanity is a parent-child mutual affection (qin) that is essentially resistant to any individualistic access (see Fan, 2002). Accordingly, Confucians cannot explore the moral probity of ESC research in terms of the total sum of individual interests (utilitarian individualism), nor in terms of the respect of individual rights (deontological individualism). This essay argues that it must be through a virtue-based, two-dimensioned ethical approach (of considering both individual and family values) that Confucians can provide an illuminating moral casuistry to shed light on the perplexing moral puzzles engendered by genetic medicine in general and ESC research in particular.

In this regard, the contrast between the Abrahamic religious concerns with early embryonic life and the Confucian thought on the issue may not be as stark as it might appear at first blush. It might have been an entire misinterpretation that Confucianism does not take early embryonic life seriously. Nevertheless, many may not feel at ease on certain conclusions that have been drawn in this essay, such as it is morally right for a Confucian family to abort the embryo in the rape case, and therapeutic cloning can be justified in the Confucian moral account. Indeed, Confucius was not Abraham. The Confucian moral symbolism cannot be exhausted by an absolutely transcendent, personal God as found in the Abrahamic religions.[20] Confucians did not find that God walked around to make covenants or contracts with humans, thus complicated moral discretion became a neat system of the Divine command, and that the only important thing left for finite humans would be ensuring their confidence in the unity of the good, the right, and the virtuous guaranteed transcendently by this Divine command system, leaving the problem of evil out of logical, intellectual and moral exploration in order to escape the horror of blasphemy. Abraham would obey every order from God. But the Confucian understanding of the mandates of Heaven is different: Heaven is more subtle and hidden than that. While Confucius started his learning at fifteen, he did not get to know the mandates of Heaven until he was fifty (*Analects* 2.4). That is, the mandates of Heaven are not a wholesale system of specific rules that have been given to the humans once and for all. They require man's own efforts to reach them. One cannot really know the mandates of Heaven unless one has cultivated virtue for a sufficiently long time. Ultimately, the mandates of Heaven are manifested in the way of Heaven that cannot be fully accomplished without perfect human virtue. And perfect human virtue, for Confucians, is irreplaceably grounded in the family love.

Indeed, there are significant differences underlying biomedical research, health care policy and bioethical explorations in the different areas of the world. This essay primarily offers an account of how Confucian family-oriented morality differs from liberal individualist morality bearing on the issues of human embryonic stem cell research. This account is part of my general project of recapturing an authentic understanding of the Confucian way of life, drawing out its implications for bioethics and health care policy, regaining a voice around the Pacific Rim. I have approached this challenge through a proposal for the reinvigoration of Confucian thought under the title of "Reconstructionist Confucianism" (Fan, 2002). Among other things, Reconstructionist Confucianism attempts to argue that it provides a more ample account of human flourishing and morality than that offered by liberal individualist moral and political theory. This essay is just one of such attempts.

NOTES

[1] A vivid intellectual confrontation between the supporters and the opponents can be found in "should federal funds be used in research on discarded embryos?" offered respectively by Myron Genel and Edmund D. Pellegrino (1999). Also see Green (2001).

[2] For a systematic explanation of the Orthodox Christian view of reproduction, cloning, abortion, and birth, see chapter 5 of Engelhardt (2000).

[3] For a general comparative study between the Christian and Confucian views on personhood, see Fan (2000).

[4] Logically, ethical theses can be independent from methodological theses. Ethical individualists do not have to be methodological individualists, and vice versa. Similarly, ethical holists do not have to be methodological holists, and vice versa. This article does not intend to explore these complex logical issues.

[5] These two fundamental liberal moral individualist principles may not have been articulated explicitly in every liberal ethicist work. But they have unquestionably underlain the moral arguments of representative contemporary liberal ethicists, such as John Rawls and Ronald Dworkin. For a clear formulation of these principles in genetic ethics, see, e.g., Buchanan, Brock, Daniels, & Wikler (2000, p. 379).

[6] In this essay, the words of value, interest, good, and worth are used interchangeably.

[7] The intrinsic worth of human individuals has long been established in the Confucian tradition. The basic idea is that, among all natural creatures, humans are invested with the most noble elements by

Heaven, and they should not be sacrificed for other creatures. This has been made clear in the five classics as well as other pre-Qin Confucian books (such as the *Analects*, *Mencius* and *Xunzi*). What is not very clear is the relation between the value of the human individual and the value of the family, which is thus the focus of this essay for exploration.

[8] Confucians have always taken the family as their primary moral community. From Confucius (551–479 BC) on, family life has been emphasized as the essential activity for human existence. For instance, when someone asked Confucius "why do you not take part in government?" the Master answered: "as the *Book of History* says, 'Oh! Simply by being a good son and friendly to his brothers a man can exert an influence upon government.' In so doing a man is, in fact, taking part in government. How can there by any question of his having actively to 'take part in government'?" (*Analects* 2: 21; translation adapted from D. C. Lau). For Mencius (372–289 BC), "there is a common expression, 'the empire, the state, the family.' The empire has its basis in the state, the state in the family..." (*Mencius* 4A: 5; translation adapted from D. C. Lau). Although the Confucian has a whole system of personal morality, including ideals of cultivating the self, regulating the family, governing the state, and making the entire world (all-under-heaven) peaceful (*Great Learning*), the two latter ideals are related indirectly to most persons, whereas the first two ideals (cultivating the self and regulating the family) are essential requirements for every individual. Finally, the moral nature and significance of the family in the Confucian tradition can be appreciated fully from its ritual system (li zhi): a series of family rituals (such as the capping, wedding, burial, and sacrificial rituals) constitute the core ceremonies and activities of Confucian lives as well as their profound moral bearings.

[9] How are individual goods defined? Basically, there are two major philosophical, incommensurable views on the issue of where the value of the individual resides. The sanctity-of-life view holds that each human life has a sacred property carrying a value or worth that is equally predicated of all human individuals, irrespective of mental or physical capacity (Ramsey, 1970). On the other hand, the quality-of-life view locates individual value in some valuable characteristics, "such as self-consciousness, rationality, the capacity to relate others, the ability to experience pleasurable states of consciousness..." (Kuhse, 1995, p. 104). For an excellent paper evaluating these two different views, see Khushf (2002).

[10] I suspect that even Aristotle may not want to have an one-way solution to the possible conflict between the good of the individual and the good of the polis. Surely for him the good of non-citizens (such as slaves) should be sacrificed. But what about free men's? In line with the good of the polis, a citizen's complete virtue should be general justice "in relation to others, not only in what concerns himself" (1129b.30); and in order to be flourishing, he "must have excellent friends" (1170b.15). However, he also identifies man's highest good is a god-like life expressing supreme virtue – focusing on theoretical study or contemplation (1176a.15). This latter "highest" individual good may not be most beneficial to the polis. Should it be sacrificed for the polis? See Aristotle (1985).

[11] Generally libertarianism holds the weak version, while contemporary liberalism holds the strong version. See Robert Nozick (1974) and Engelhardt (1996). However, in debating with communitarians, liberals sometimes claim that they only hold the weak version. See, e.g., Kymlicka (2002, p. 223). In any case this controversy does not affect the argument employed in this paper.

[12] Small children have not yet fully developed their self-determination capacity and thereby are not competent to make contracts with their parents regarding ways in which they should be raised. Liberals have to make decisions for their children. Unlike devout Christians who will make their children Christians and serious Confucians who will train their children to become filial, liberals have to promote their children's capacity for self-determination: for liberals, it is essential that individuals be able to choose, reflect on, and revise their life plans by themselves; namely, they must practice self-determination. Accordingly, the liberal ideal is not to embed their children in a way of life that accords with their parents' life plans or views of the good. Rather, they bear the moral obligation to bring up their children as persons exercising self-determination. Only in this way, liberals would argue, have their children been treated as ends (although they are not actually ends yet in the sense of moral agents), not merely as means to their parents' ends.

A typical liberal view in this regards is as follows:
A child's good is more fully determined by the developmental needs of children

generally at that age than by his or her current but predictably transient goals and preferences. These developmental needs are based in significant part on the aim of preparing the child with the opportunities and capacities for judgment and choice necessary for exercising self-determination as an adult. Consequently, efforts to promote children's well-being focus prominently on fostering these abilities and opportunities so that as adults they will be able to choose, revise over time, and pursue their own particular plans of life, or aims and values, now suited to the adults they have become (Buchanan & Brock, 1989, pp. 227–228).

[13]   Still, it is logically clear that if embryos are only potential persons, they do not have the rights of persons (Engelhardt, 1996, p. 142).

[14]   See Singer (1983, 1993); Kuhse (1987, 1991). For an excellent critique of the utilitarian view, see Khushf (2002).

[15]   See a helpful comment offered in Chan (1963, p. 44).

[16]   Therapeutic cloning is morally quite different from reproductive cloning. Here the purpose is not to recreate a human being, but offer stem cells for treatment. Because the stem cells thus harvested are immunologically compatible with the patient, such cloning as a therapeutic intervention is extremely technically promising. For a Confucian account of reproductive cloning, see Fan (1998).

[17]   All these assumptions require Confucian family-value-based, sophisticated arguments that go beyond the scope of this essay.

[18]   Similarly, it should be morally permissible to use the tissues from an embryo that has been aborted, insofar as the abortion is morally justified.

[19]   I am not sure how seriously this would affect the development of ESC research. It is controversial whether stems cells abstracted from bone marrows are equally promising.

[20]   For a helpful account of Heaven under the understanding of Confucius and Mencius, see Ivanhoe (2006).

REFERENCES

Aristotle (1958). *The Politics of Aristotle*, E. Barker (trans.), Oxford University Press, Oxford.

Aristotle (1985). *Nicomachean Ethics*, T. Irwin (trans.), Hacket Publishing Company, Indianapolis.

Brody, B. (1972). 'Thomson on abortion,' *Philosophy & Public Affairs*, 1 (Spring 1972), 335–340.

Buchanan, A. & Brock, D.W. (1989). *Deciding for Others: The Ethics of Surrogate Decision Making*, Cambridge University Press, Cambridge.

Buchanan, A., Brock, D.W., Daniels, N., & Wikler, D. (2000). *From Chance to Choice: Genetics and Justice*, Cambridge University Press, Cambridge.

Chan, W.T. (1963). *A Source Book in Chinese Philosophy*, Princeton University Press, Princeton.

Confucius (1971). *Confucian Analects, The Great Learning & The Doctrine of the Mean*, James Legge (trans.), Dover Publications Inc., New York.

Confucius (1979). *Analects*, D.C. Lau (trans.), Penguin Books, Suffolk.

Engelhardt, H.T., Jr. (1996). *The Foundations of Bioethics*, Oxford University Press, Oxford.

Engelhardt, H.T., Jr. (2000). *The Foundations of Christian Bioethics*, Swets & Zeitlinger Publishers, Lisse.

Fan, R. (1998). 'Human cloning and human dignity: Pluralist society and the Confucian moral tradition,' *Chinese & International Philosophy of Medicine*, I: 3, 73–93.

Fan, R. (2000). 'Can we have a general conception of personhood in bioethics?,' in G. Becker (Ed.), *The Moral Status of Persons: Perspectives on Bioethics* (pp. 15–27), Rodopi, Atlanta.

Fan, R. (2002). 'Reconsidering surrogate decision making: Aristotelianism and Confucianism on Ideal Human Relations,' *Philosophy East & West*, 52, 346–372.

Fan, R. (2002). 'Reconstructionist Confucianism and bioethics: A note on moral difference,' in H.T. Engelhardt, Jr. and L. Rasmussen (Eds.), *Bioethics and Moral Content: National Traditions of Health Care Morality* (pp. 181–187), Kluwer Academic Publishers, Dordrecht.

Genel, M. & Pellegrino, E. (1999). 'Should federal funds be used in research on discarded embryos?' in *Physician's Week*, Vol. XVI, No.37 (October 4, 1999).

Green, R. (2001). *The Human Embryo Research Debates: Bioethics in the Vortex of Controversy*, Oxford University Press, Oxford.

Ivanhoe, P. (2006). 'Heaven as source for Ethical Warranty in Early Confucianism,' *Dao: A Comparative Philosophy*, 6.1 (forthcoming).

Khushf, G. (2002). 'The domain of parental discretion in treatment of neonates: Beyond the impasse between a sanctity-of-life and quality-of-life ethic,' in J. Tao (Ed.), *Cross-Cultural Perspectives on the (Im)Possibility of Global Bioethics* (pp. 277–298), Kluwer Academic Publishers, Dordrecht.

Kuhse, H. (1987). *The Sanctity-of-Life Doctrine in Medicine: A Critique*, Oxford University Press, New York.

Kuhse, H. (1991). 'Severely disabled infants: Sanctity of life or quality of life?,' *Baillieres Clinical Obstetrics and Genecology*, 5(3), 743–759.

Kymlicka, W. (2002). *Contemporary Political Philosophy: An Introduction*, 2nd Edition, Oxford University Press, Oxford.

Mencius (1970). *Mencius,* D.C. Lau (trans.), Penguin Books, Suffolk.

Nozick, R. (1976). *Anarchy, State, and Utopia*, Basic Books, New York.

Ramsey, P. (1970). *The Patient as Person*, Yale University Press, New Haven.

Singer, P. (1983). 'Sanctity of life or quality of life,' *Pediatrics*, 72 (1), 128–129.

Singer, P. (1993). *Practical Ethics*, 2nd Edition, Cambridge University Press, Cambridge.

Thomson, J. (1971). 'A defense of abortion,' *Philosophy & Public Affairs*, 1 (Fall 1971), 47–66.

Xunzi (2003). *Xunzi*, B. Watson (trans.), Columbia University Press.

CHAPTER 11

A CONFUCIAN EVALUATION OF EMBRYONIC STEM CELL RESEARCH AND THE MORAL STATUS OF HUMAN EMBRYOS*

SHUI CHUEN LEE

National Central University, Taiwan

One of the post human genome challenges for geneticists is to find out the mechanism of genes and this is best investigated through genetic experiments with undifferentiated stem cells. The most desirable type is called pluri-potent stem cells, which are usually retrieved from an embryo in its early stages.[1] They are cells that will eventually develop into all the tissues and organs contained in the human body, and thus are the most suitable for such research. While there are a few types of stem cells in an adult human, they are usually in scarce amounts and are difficult to obtain, and sometimes the process may involve great risk to the subject. What is most important is that stem cells are more or less somewhat differentiated and are already destined to grow into certain types of tissues. Though there are certain reports that some types of stem cells can be re-vitalized as undifferentiated stem cells or pluri-potent stem cells, the prospect does not seem very good and it is doubtful whether they are as truly pluri-potent as those that appear in the early stages of an embryo. The same poor prospect applies to stem cells obtainable from umbilical cords of a newborn baby.

Now, the subjects or the biological material used for these types of experiments, namely, the stem cells or pluri-potent stem cells, pose no ethical problem, except for being subject to the usual type of regulations for they are just a group of cells and not some subject of life that can be harmed. However, to obtain embryonic stem cells for experimentation, the subject embryo will be dismembered and destroyed in the process. This, however, may be morally problematic. Using a living human being purely for the sake of experimentation and causing its destruction seems to

---

* This essay is one of the papers produced by my project on "The Personhood and Moral Status of Embryos: The Challenge of Human Genome", which is a three-year project (2002–2005) supported by the National Science Council, Taiwan.

149

violate some of our basic beliefs, such as the beliefs that one should not treat human beings instrumentally and that one should not treat subjects of experiments without due concerns for their benefits and protection. However, the benefits that can be achieved through stem cell research are so great that one needs exceedingly good reasons for not doing so. In general, Confucianism supports scientific research that aims to obtain more and better knowledge of ourselves and the world at large. Consequently, Confucianism endorses stem cell research. However, when it comes to using and destroying embryos in order to carry out such experiments, Confucians become more cautious. Whether such actions are morally permissible depends on the moral status of embryos involved, which we shall poke deeper into before coming to any solid conclusions.

I. EMBRYOS, PERSONHOOD AND MORAL STATUS: WARREN'S MULTI-CRITERIA THEORY

One way of approaching the problem is to argue for or against the notion that an embryo as an entity with full rights as a person. Now, if we accept the basic idea that an entity is not an individual if it does not possess some sort of self-identity and thus there is no reason to attribute to it the status of a person, then an embryo cannot be a person and cannot claim the rights of a person. In the case that we are considering, the embryo for ES cells is usually 14 days old or less and before the primitive streak appears. It does not have a workable organ needed to maintain memory and self-identity. As such, it is not a person in the usual sense. Some may like to regard it as a potential person. However, the theoretical and practical problems of treating potential persons, such as embryos and fetuses, as persons are well-known and insurmountable, and I presume that it is justified to refuse to confer full personhood to an embryo.[2] Thus, the notion of personhood is not available for us to deal with in our present problems.

However, even if the embryo is not a person, this does not mean that an embryo has no moral weight that can counter our arbitrary treatment of it. How much moral weight an embryo has turns on its moral status. By moral status, I would like to refer to Mary Ann Warren's delineation of this concept as follows:

> To have moral status is to be morally considerable, or to have moral standing. It is to be an entity towards which moral agents have, or can have, moral obligations. If an entity has moral status, then we may not treat it in just any way we please; we are morally obliged to give weight in our deliberations to its needs, interests, or well-being. Furthermore, we are morally obliged to do this not merely because protecting it may benefit ourselves or other persons, but because its needs have moral importance in their own right (Warren, 1997, p.3).

Basically, if an entity is considered a person or possesses personhood, then it has the same moral status as a person, which usually means it has the highest moral status rank and has rights to life and liberty. Those who take a zygote to be a person, claim that it has the same moral status as a normal adult. This same moral status is also offered to a human being further developed than a zygote, such as an embryo or a fetus. Thus, killing an embryo or a fetus would be the same as killing an innocent person. Hence, according to this view, derivation of pluri-potent stem cells from an embryo by killing it is abominable. However, as was stated in the last paragraph, because we do not have enough reason to confer to an embryo the status of personhood, the moral standing of an embryo cannot be equal to a normal adult person.

Now, we may consider attributing to embryos a moral status less than that of a full person. Since Warren gives a very excellent and detailed theory of moral status, we may start with her evaluation of an embryo's moral status. After a lengthy discussion of a number of famous single criteria accounts of moral status, Warren disposes them off as inadequate and proposes her multi-criteria account which utilizes seven principles (Warren, 1997). Warren's views in terms of the problem of abortion can be easily extended to embryos:

> There are two weaknesses inherent in approaches to the ethics of abortion that focus exclusively upon the nature of the fetus, and the moral status this is thought to imply. In the first place, the moral status of embryos and fetuses cannot be determined solely through a consideration of their intrinsic properties, as most of the uni-criterial approaches require. Their unique relational properties are also relevant; in particular, their location within and complete physiological dependence upon the body of a human being who is (usually) both sentient and a moral agent. These relational properties have to be considered in determining the moral status that may reasonably be ascribed to fetuses (Warren, 1997, pp. 201–202).

Warren's discussion of fetuses is equally applicable to embryos. As the second weakness has to do with women's right to abortion, which is not our concern here, I shall skip it. However, the relation that bears some significant importance for the status of embryos is not as clear as we would like. If the carrying mother wants to have a child, no one can override her. If she decides to have an abortion, the moral status of embryo seems easily overridden by the autonomy and rights of a sentient moral agent. What is significant for us is the moral status of embryos *in vitro*, rather than those *in vivo*, for we are talking about embryos for the development of cell lines.

Warren then employs her multi-criteria account to explain the moral status of embryos. Her discussion is once again in terms of the ethics of abortion:

> Fetuses at this stage [i.e. first ten weeks] are obviously not moral agents and thus are not accorded full moral status by the Agent's Rights principle. Nor, at this stage, are they capable of sentience; thus neither the Anti-Cruelty principle nor the Human Rights principle apply. They are, however, alive, and thus have moral status based upon the Respect for Life principle. Since they are regarded by some people as having full moral status, they may gain some moral status through the Transitivity of Respect principle. However, unless they have full moral status for some other reason, the status that they can be accorded on the basis of the Transitivity of Respect principle is limited by the moral rights that women enjoy under the Agent's rights principle. The question, then, is whether there is any independent reason to accord full moral status to zygotes, embryos, and fetuses (Warren, 1997, pp. 202–203).

Here Warren takes the relation of embryos to other moral agents into account in her Transitivity of Respect principle. It is clear that the moral status of embryos enhanced by the Transitivity principle seems to be rather minor, especially in comparison with the important human rights of a woman, who is a sentient moral agent. It is also clear that for Warren, an embryo does not have any right to life, as it is not sentient and thus very far from having any human rights by itself. However, it is not clear how much moral weight we can accord to them when it comes to using embryos for the derivation of cell lines for research. It seems that the importance and benefits of stem cell experimentation outweigh any moral consideration of the status of embryos.

However, if one party debating this issue takes an embryo to be a living thing with the same rank as a moral agent, killing an embryo is a violation of the right to life of a moral agent and is detrimental to our moral community. Thus, for anyone who confers such a moral status to embryos, it is unacceptable to use and kill embryos in experiments. On the other hand, the opposing party will not accept such a conferment and limits the status of the embryo to something that has the same status as any non-sentient living thing. Thus, for this party, doing research with embryos is by no means wrong. This issue between the opposing views seems irreconcilable. Unfortunately in this case, the two sides cannot be left to do as they wish within their own circle of rights as it involves a public policy where both sides cannot be exempted. It seems fairly clear that Warren would claim that the side advocating the protection of embryos violates the prerequisite of fulfilling the requirements of the other six principles, particularly the third one, the Agent's Rights principle. For, as we said earlier, an embryo does not have any substantial organs for feelings and it cannot be sentient in our usual sense of the word. Because it is not a sentient entity, an embryo cannot have a status comparable to a person or a moral agent. Hence, such a conferment is unreasonable and void of making the claim to be respected. The opposing party, who rejects the claim that embryos are sentient, has an advantage in that its judgment of the moral status of embryos is consistent with the other six principles and thus its claims need to be respected by the party arguing for the protection of embryos. Because an embryo cannot be sentient in its first fourteen days of life, respect must be given to the judgment that the moral status of an embryo is the same as any other non-sentient being and thus experimenting on them is fully justifiable. However, the strong objections of each party need to be seriously taken into account as we are all in some sort of community, which is the locus of the working of Warren's seven principles. Furthermore, Warren and others seem to assume that all embryos are ontologically the same and thus when we are allowed to experiment on one type of embryo, this implies we are allowed to experiment on all other types of embryos. However, the legitimacy of the use of different types of embryos must also be debated, in order to produce a more complete response to the issues at hand.

II. MORAL COMMUNITY VERSUS SOCIAL COMMUNITY

In order for Warren's theory to work, we cannot merely subscribe to the notion that we are a community of unrelated individuals. Rather we need to subscribe to the notion that we belong to a moral community which binds its members morally. If people who hold opposing moral views are regarded as being two different communities of moral friends living together in a secular society of moral strangers (Engelhardt, 1996), the solution cannot be convincing for one or the other of the two parties. They will be in a deadlock and the dispute cannot be settled peacefully in the sense that permission must be sought for doing or prohibiting embryonic stem cell experimentation. As a public issue, consensus is needed.

In view of the fact that we have such rich moral cultures and even such diverse opinions on moral matters, we are the only species that care for the rightness or wrongness of our acts. Homo sapiens are moral beings. Morality is at least one essential aspect of our being. Thus, we are not just living together socially, but are also mingling with each other ethically or morally. In other words, we are all in a moral community. The moral community is part and parcel of human society and that means we have stronger moral commitments to each other than just a principle of permission.[3] The moral community is a concept lurking behind almost all of our moral problems. However, it is usually invoked without much explanation of what it means and what moral signification it bears upon our moral judgments. For example, although Warren agrees that moral agents cannot be social atoms (Warren, 1997, p. 158), she often speaks about the social community or human community without explicitly referring to its moral character. Even when she refers to the moral community, she regards it as "the set of entities to which we ascribe full and equal moral status" (Warren, 1997, p. 15), as if moral community is just a name for a group of independent or atomistic individuals. If this is how human beings form socially, relations between moral agents cannot be of any moral significance and some of her principles would become vacuous.

Warren agrees with Hume's belief that our "moral feeling," that is, care about other's sufferings, is the foundation of all human morality (Warren, 1997, p. 12). According to Confucianism, this is our primordial moral consciousness - our unbearable mind of others' suffering. It reveals itself when we feel or know some living entity has incurred or is incurring a serious injury, whether we have any relationship with this living entity or not. Not only do we care about unnecessary harm that living things incur, but we also care about harms that are permitted if they eschew our sympathetic feelings. This sense of moral commitment to each other makes us a moral community. The basic members of this moral community are moral agents, whose characteristics are those underlined by the Kantian or Confucian conception of humanity within us. Broadly speaking, moral agents are beings that are self-conscious, rational, free, and have minimal moral sense (Engelhardt, 1996). The central constitutional element of this moral community is our sense of moral commitment to each other. When we say human beings have moral sense, we are referring to our compassion for the sufferings of other sentient beings, human and non-human. When we say we are free, we mean that we can act competently and independently within a morally connected community. In other words, moral agents are embedded in some moral relationship with each other. As Confucians, we take seriously the interrelatedness of the members of the moral community and give more weight to the real worries and objections of our fellow members. We have a natural passion for our own offspring and freely assume duties to them, and at the very least have moderate benevolence towards our fellow man as Hume observed.[4] This care and relational concern gives human society its moral character and its related concept of moral community gives the moral foundation for Warren's principles.[5]

Embryos bear a specific relation to us, which greatly increases their moral standing. For embryos bear a much closer relationship to members of our moral community than others and arouse some deep concerns among us. That explains why the answer cannot be so straight forward as Warren's principles imply. Furthermore, under such a moral network, greater respect should be given to the moral standing of embryos by those who regard embryos as having the same moral status as other non-sentient entities. This implies that except when there is no other way possible, we should try to use sources other than embryos to derive stem cells. However, embryos cannot enjoy the same moral status as children in our moral community. We have no reason to claim that scientists working on stem cell research are killing innocent persons. However, the idea of a moral community enhances the moral status of embryos considerably and commands that all parties be cautious and morally prudent when it comes to using embryos. With such moral excuses, we can now scrutinize in more detail the different ways that embryonic stem cells are obtained for experimentation.

III. MORAL STATUS OF VARIOUS TYPES OF EMBRYOS FOR EXPERIMENTATION

Embryonic stem cells can be obtained from four different sources. They can be obtained from the remains of abortions, from extra embryos created for artificial reproduction, from embryos created by somatic cell nuclear transfer, or from embryos specially produced for research from donated gametes.

Apart from the moral problems of abortion, the remains of a fetus can be viewed as a biologically dead body. Being dead, the fetus is not a member of our moral community nor can it be enjoined by the seven principles that Warren speaks of. Its legal guardian has the right to donate it for research as is the case with cadaver donations in general. In the case of extra embryos, that is those embryos created for IVF and that will no longer be used for this purpose thus they will be destroyed after a number of years, their significance and importance for stem cell experimentation are considerable, although such embryos have considerable moral standing. Such embryos are the perfect source for stem cell experimentation since they were not created for this purpose and thus, do not involve researchers in the debate of creating embryos for experimentation. Since such embryos are destined to be destroyed, and assuming that the opposing parties accept this fact, it seems permissible to use these embryos for the morally prudent purpose of saving lives and relieving the suffering of others and future generations.

As for the creation of embryos solely for such experiments, there is the stronger objection that it makes human life a mere instrument. In view of the considerable moral weight that is added onto the intrinsic properties of an embryo, the infringement or abuse of our respect for the related members of our moral community makes it impermissible without stronger counter considerations. Some might argue that since the ontological status of embryos is the same, if we allow one type of embryo to be experimented upon, we have no reason to prohibit the other types from being used for the same purpose.[6] However, the accentuated moral status of embryos that results from their interaction with members of the moral community, in some sense commands that we pay more respect to such human life and that we need to take their interests as one of our primary concerns. But it can also be argued that embryos left over from IVF treatments are originally created for their own benefit. When this falters, the next best use is to use to them to benefit others. Confucians have no qualms with employing the remains of fetuses and extra embryos for scientific research. It is somewhat like the best outlet in the aftermath of a tragedy. However, creating them solely for experiments is quite different from the act of procreation. It comes closer to creating a tragedy for some good reason, which at least initially does not seem morally prudent if there are other outlets. Although embryos are not yet persons nor sentient, Confucianism asserts that they are a kind of human life, which needs to be respected and rejects making a tragedy of any life if there are other outlets to avoid so. Hence, creating embryos for the purpose of research is unacceptable and morally imprudent especially in view of the strong emotional objections from the other parties. Furthermore, even those who regard all embryos as being ontologically the same, they give at least some minimal moral status to them and the destroying of an embryo is at least less morally desirable than not destroying any of them. In other words, unless there is no other choice, such as using extra embryos for a significant research designated for the study of the relief of pain felt by patients now and in the future, we have to abstain from creating them and destroying them if we could create the same research by other means.

On the other hand, creating special embryos from the gametes of those who have genetic diseases for stem cell experimentation becomes a question of justice. It comes close to being a clash between the moral status of embryos and the rights of certain groups of moral agents. It is unreasonable to prohibit their procreation, nor let their progeny carry on the diseases when there are chances for cure. Relieving the pain of others is of paramount value for Confucians. Such pain-relieving action carries considerable moral weight and outweighs the moral standing of embryos. The cure of genetic diseases lends support for such research as it may be the best means to end the prevailing tragedies that these ethic groups or families have been facing for generations. The specific goals of therapeutic cloning researches prevail as well.[7] Whether the embryos in such cases are obtained through SCNT from donated gametes or not does not make any moral difference.

IV. CONCLUDING REMARKS

The moral status of embryos has long been scrutinized because of the problem of abortion. Its discussion has been brought to the front stage again by the emergence of embryonic research. I have attempted to present my case with the help of Mary Anne Warren's multi-criteria theory of moral status in order to argue why embryonic stem cell experimentation is by all means morally permissible – although she may not agree with all that I have said. I have integrated the concept of moral community into her theory. This leads to the conclusion that creating embryos, whether by donated gametes or by SCNT, solely for experimentation is not acceptable except in some special cases where genetic diseases are involved that cannot be investigated in other ways that are less morally objectionable.

Warren's theory is a good starting point. It is very helpful in clarifying what elements we need to consider when deciding whether stem cell experimentation is morally permissible and gives us a model for which we may assess the relative merits of such experimentation. In fact, we have adopted most of her work. Her theory suffers somewhat, such as in the aspect of principlism, in that it lacks a grounded and coherent account for such criteria. The notion of a moral community is an attempt to provide the grounding for her theory. However, this only takes us half way towards a full grounding of her theory. The moral priority of the notion of moral community over social community and the moral significance of the notion of harm over autonomy are also part of the framework of this grounding problem. The exploration of these issues requires more groundwork than we can cover in this paper.

NOTES

[1]   There are in fact two types of embryonic stem cells. The first type is derived from the inner lining cells of a blastocyst and called embryonic stem cells (ES). The second type is derived from the primordial germ cells of an embryo and called embryonic germ cells (EG). The later usually involves the dead body of an abortion. Though both involve the death of an embryo, they have a slightly different ethical problem in that the former involves the killing of an embryo for its ES cells, while the latter only acquires the EG cells after a fetus is being aborted. The following has the former type in mind while the latter type will be dealt with towards the end of the paper. Cf. *Ethical Issues in Human Stem Cell Research* (1999).

[2]   There are numerous texts and articles on this problem, especially the part it touches abortion. For a concise and to the point presentation that support my assertion, please refer to Mary Anne Warren's book, *Moral Status: Obligations to Persons and Other Living Things* (1997).

[3]   I presented my argument for in more detail in my paper ' The Possibility of a Substantial Secular Bioethics: Towards a Confucian Bioethics,' which was presented in The Second International Conference of Bioethics, in 1998, Chungli, Taiwan.

[4]   That sentient animals have similar caring attitude towards their offspring seems quite common. While, human being may be the few species that show gratefulness towards their parents, which is taken by Confucians as an important element for being human.

[5] To work out the details need another fairly long essay and have to be left for another occasion. The following discussion on the moral status of embryos may convey some of my ideas in this respect.

[6] This kind of argument is proposed by John Harris, Peter Singer and others. For more details, please be referred to John Harris's *Wonderwoman and Superman: The Ethics of Human Biotechnology.*

[7] I shall leave the case of reproductive cloning and the general problem of human cloning for another occasion.

REFERENCES

Engelhardt, H.T., Jr. (1996). *Foundations of Bioethics*, 2nd Ed., Oxford University Press, New York.

Harris, J. (1992). *Wonderwoman and Superman: The Ethics of Human Biotechnology*, Chapter 2, Oxford University Press, Oxford.

National Bioethics Advisory Commission (1999, September), *Ethical Issues in Human Stem Cell Research*, Vol.1, Esp. Chapter 2, Rockville, Maryland.

Warren, M.A. (1997). *Moral Status: Obligations to Persons and Other Living Things*, Chapter 9, Clarendon Press, Oxford.

CHAPTER 12

REGULATIONS FOR HUMAN EMBRYONIC STEM CELL RESEARCH IN EAST ASIAN COUNTRIES

A Confucian Critique

HON CHUNG WONG

National Central University, Taiwan

I. INTRODUCTION

Human embryonic stem cell (HESC) experimentation is regarded as being one of the most important fields in biomedical research. Such experimentation aims to discover how stem cells differentiate into more specific cells in the human body. Moreover, these stem cells will be important tools in the development of life-saving drugs and cell-replacement therapies designed to treat injuries or diseases such as Alzheimer's disease, Parkinson's disease, heart disease and kidney failure. However, how these cells are derived raises various ethical and legal problems. Presently, stem cells are derived from four sources: cadaveric fetal tissue, embryos remaining after infertility treatments, embryos made solely for research purposes using IVF, and embryos made using somatic cell nuclear transfer into oocytes (SCNT). Deriving stem cells from cadaveric fetal tissue is controversial because of the ethical problems related to abortion. Deriving stem cells from the other three sources is controversial because the subject embryo will be destroyed in the process. Treating embryos in this way means that we regard it as a means to our other ends rather than as an end in itself, which violates the moral belief that a human being should not be treated instrumentally. Moreover, each of the last three kinds of derivation has its own ethical problems and should be considered independently. It may be case that the derivation of stem cells from research embryos and through cloning embryos may be morally acceptable only in very limited cases. Hence, numerous nations and states in the world have made laws or guidelines to regulate human embryonic stem cell research and the derivation of stem cells from different sources.

Many scientists in East Asian countries, such as China, Taiwan, Singapore, Japan and the Republic of Korea are engaged in HESC research. If we examine in

159

S.C. Lee (ed.), The Family, Medical Decision-Making, and Biotechnology, 159–166.
© 2007 *Springer.*

detail the relevant policies adopted by these countries, we find that regarding the different sources of stem cell derivation, these policies are more liberal than those adopted by some western countries such as the United States and most European countries except the United Kingdom and Belgium. This may reflect their cultural and religious differences. On the other hand, this may also reflect the fact that governments in these countries have not seriously based their policies on their own core ethical values. Economic, research and therapeutic inducements have dominated the process of policymaking. Indeed, the cultures and societies of these countries are more or less influenced by Confucian thinking. It may serve as a touchstone to test whether the policies on human embryonic stem cell research formulated by these governments reflect the core ethical values in these countries.

In this paper, I begin by distinguishing among six possible policy options regarding HESC research, and then report the different policies adopted by the East Asian countries mentioned above. I also compare the rationale of these policies with those adopted by some western countries, such as the United States and the United Kingdom. Finally, I review these policies from a Confucian perspective.

II. THE REGULATIONS FOR HESC RESEARCH IN EAST ASIAN COUNTRIES

According to LeRoy Walters, there are six possible policy options regarding human embryo research and HESC research (Walters, 2004):

Option 1: No human embryo research is permitted, and no explicit permission is given to perform research on existing human embryonic stem cells.

Option 2: Research is permitted only on existing human embryonic stem cell lines, not on human embryos.

Option 3: Research is permitted only on remaining embryos no longer needed for reproduction.

Option 4: Research is permitted both on remaining embryos (see Option 3) and on embryos created specifically for research purposes through in vitro fertilization (IVF).

Option 5: Research is permitted both on remaining embryos (see Option 3) and on embryos created specifically for research purposes through somatic cell nuclear transfer into human eggs or zygotes.

Option 6: Research is permitted both on remaining embryos (see Option 3) and on embryos created specifically for research purposes through the transfer of human somatic cell nuclei into nonhuman animal eggs, such as rabbit eggs, for example.

2.1 China

In October 2001, the Ethics Committee of the Chinese National Human Genome Center at Shanghai formulated a set of proposed guidelines for HESC research in China. The Guidelines state that the transfer of human nuclei into nonhuman, mammalian eggs should be permitted:

> Use of the "human-animal" cell fusion technique is permissible in basic research with non-clinical application if the requirements expressed in the first three points of the present Article 14 are satisfied. However, use of the product formed by combining a human somatic cell nucleus with the oocyte cytoplasm of an animal using the "human-animal" cell fusion technique is strictly forbidden in therapeutic cloning research for clinical application in the treatment of human diseases (Ethical Committee of the Chinese National Human Genome Center at Shanghai, 2002).

The guidelines also state that the derivation of ES cells from cloned embryos should be permitted, since the problem of incompatibility can be solved if the cloned embryo is derived using the material from the somatic cell of a patient (SCNT) (Ethical Committee of the Chinese National Human Genome Center at Shanghai, 2002).

During the same year, an interdisciplinary group of scientists and ethicists in Beijing formulated another set of proposed guidelines. The guidelines state that combining human gametes with animal gametes is strictly forbidden, but research on combining human somatic-cell nuclei with animals is permitted under close supervision. The guidelines also mention that using IVF to create embryos with donated gametes and using SCNT to create embryos with donated somatic cells and eggs can be permitted under strict and reasonable conditions and under the principle of donors' informed consent.

In January 2004, the Ethical Guiding Principle on Human Embryonic Stem Cell Research was issued by Ministry of Science and Technology and Ministry of Health in China. These guidelines state that human embryonic stem cells used for research purposes can only be derived from the following sources so long as voluntary agreement is secured: (1) spared gamete or embryos after *in vitro* fertilization (IVF); (2) fetal cells from accidental spontaneous or voluntarily selected abortions; (3) embryos obtained by somatic cell nuclear transfer technology or parthenogenetic split embryos; (4) germ cells voluntarily donated. The guidelines also forbid hybridizing human germ cells with germ cells of any other species (Ministry of Science of Technology and Ministry of Health, 2004).

2.2 Taiwan

After a series of three public hearings held by Taiwanese Society for Reproductive Medicine, the Department of Health in Taiwan formulated guidelines for the conduct of HESC research in February 2002. According to the guidelines, surplus embryos that are no longer needed for reproduction may be used for stem cell research. However, donor eggs and sperm are not to be used to create embryos for research purposes. Regarding research cloning, the Department deferred a decision, saying:

"The issue of producing human embryos for research purposes through nuclear transfer should be examined in greater detail because in this case multiple realms are involved" (Department of Health, 2002).

In June and September 2003, two groups of congressional representatives presented two bills concerning "Research and Protection for Human Embryos and Embryonic Stem Cells" to the Taiwanese Parliament respectively (Legislative Yuan, Republic of China, 2003). One of the bills stated that the derivation of ES cells from cloned embryos should be permitted, since ES cells derived from other sources may not be enough for experiments. Two bills have been transferred to the Committee of Science and Technology of the Parliament for further examination.

2.3 Singapore

In 2000, the Cabinet of Singapore government appointed the Bioethics Advisory Committee (BAC) to examine the ethical, legal and social issues arising from biomedical research in Singapore. In particular, the Human Stem Cell Research Sub-Committee was formed under the BAC in 2001 to specifically deal with the relevant issues arising from HESC research. BAC finished its report in June 2002 (Bioethics Advisory Committee, 2002). In November 2003, the Minister of Health drafted the Regulation of Biomedical Research Bill and placed the bill on the web for further public consultation (Ministry of Health, Singapore). Both the BAC report and the draft law accept the creation of embryos for research purposes through IVF (Option 4) and through SCNT (Option 5). The BAC report stated that the creation of research embryos can only be justified where (1) there is strong scientific merit in and potential medical benefit from such research, (2) no acceptable alternative exists, and (3) on a highly selective, case by case basis, with specific approval from the proposed statutory body. Moreover, the report mentioned that reason to support the derivation of ES cells from cloned embryos is that tissues repaired by such ES cells would be more likely to be immunologically compatible with the intended recipient. Therapeutic cloning also enables scientists to learn about the mechanisms of reprogramming adult cells to behave like stem cells, and potentially making it unnecessary to resort to using embryos as a source of stem cells.

2.4 South Korea

In late December 2003, the National Assembly of South Korea passed the Bioethics and Biosafety Bill. This law prohibits reproductive cloning and routinely permits the use of remaining embryos that have been frozen for at least five years for research (Option 3). The South Korean government will also allow limited research involving nuclear transfer (Option 5), under guidelines to be developed by a newly created National Ethics Committee. Research cloning experiments will require the formal approval of the South Korean President (Walters, 2004).

2.5 Japan

In Japan, research guidelines for HESC research were drawn up by an advisory committee for the Ministry of Education, Science and Technology, and were released in September 2001. These guidelines permit only research with remaining embryos that have been produced through IVF (Option 3) (Walters, 2004).

After three years of debate, the bioethics committee of the government's Council for Science and Technology Policy permitted the production of cloned human embryos for basic research only, with a set of stringent conditions on June 20, 2004.

III. THE SIMILARITIES AND DIFFERENCES ON REGULATIONS FOR HESC RESEARCH BETWEEN EAST ASIAN COUNTRIES AND WESTERN COUNTRIES

If we compare the regulations for HESC research in East Asian countries with those in most of the European countries and in the United States, we will find that the latter is more conservative than the former. Within Europe, several countries have not expressly permitted any kind of HESC research (Option 1), such as Austria,[1] Ireland[2] and Norway (European Commission, 2004) but only in Norway is any kind of HESC research expressly forbidden. Moreover, Germany has adopted Option 2. Germany permits the importation and use of human embryonic stem cells that were derived outside of Germany before January 1, 2002, and a majority of European nations have adopted Option 3, such as The Czech Republic, Denmark, Finland, Greece, Hungary, the Netherlands, Russia and Spain. Only the United Kingdom and Belgium have accepted both Option 4 and 5 (Walters, 2004).

In the United States, the federal government has only established a funding policy for HESC research, and allows the state governments to decide which kind of such research is to be permitted. Hence, President George W. Bush presented a speech on August 9, 2001 to endorse Option 2, with a time limit on the date by which the HESC lines must have been created. On the other hand, different states have adopted their own policy options about HESC research, ranging from prohibiting all HESC research to permitting research on embryos created through SCNT into human eggs or zygotes (Walters, 2004).

The reason that some of the countries mentioned above have adopted policies that prohibit or do not yet permit any kind of HESC research may be explained by their views about the moral status of embryos, which are influenced by Christianity. According to the Christian view, an embryo is an entity with the full rights of a person. However, the generally accepted time range for an embryo being used for HESC research is within 14 days after conception, as the primitive streak has not appeared in the embryo at that time. This means that the embryo has not developed its own nervous system. Therefore, it seems hardly appropriate to say that the embryo has its own memory and self-identity. Since memory and self-identity are the essential criteria for attributing an entity the status of a person, then the embryo should not be regarded as person. Moreover, we cannot argue for the personhood of an embryo by appealing to the potentiality argument, since it is usually regarded as problematic. The fact that an embryo has the potential to qualify as a person, if certain conditions are met, does not confer to it the rights that belong to a person until those conditions are met. On the other hand, because these countries also permit assisted reproduction through IVF, there will be a certain number of embryos left after infertility treatments. If those embryos will inevitably be destroyed after a certain number of years, it may be a better outlet to donate them for HESC research.

The policy options for HESC research adopted by the federal governments in Germany and the United States seem to be inconsistent. If they allow research on existing human embryonic stem cell lines, which are obtained from remaining embryos no longer needed for reproduction and result in the subject embryo's death, why cannot they go further and allow research on remaining embryos without a time limit? The defender of this kind of policy may argue that more embryos will die if we accept the latter option. However, the embryos in this case will be destroyed after a period of time, whether they are donated to research or not, so we might think that we ought to use them for HESC research in order to relieve the suffering of others and future generations. Moreover, a report of the President's Council on Bioethics in the United States has argued that President Bush did not adopt this policy because of his views about the moral standing of embryo; rather he adopted this policy because he is uncertain about this issue (President's Council of Bioethics, 2004). If that was the case, then President Bush should defer making a decision rather than hastily adopting a policy whose moral implications are unclear to him. Hence, we may regard this kind of policy as a result of compromise between the pros and cons of HESC research.

According to the guidelines established by the British Government, all kinds of HESC research may be accepted, except research on embryos created specifically for research purposes through the transfer of human somatic cell nuclei into nonhuman animal eggs (House of Lords, 2004). It is also the case in many East Asian countries, which may be explained by the fact that embryos are not viewed as a person in both regions. However, in Britain, some kinds of research, especially the creation of embryos only for research purpose are involved, should be conducted under strict restrictions. Are there any corresponding restrictions, which should be applied in the cases of East Asian countries? As for the creation of embryos solely for HESC research, some might argue that since the ontological status of embryos is the same, if we allow one type, we have no reason to prohibit the other types. However, according to the Confucian view, the accentuated moral status of embryos through their interaction with society in some sense commands that we pay more respect to such human life forms and need to take their own interests as one of our primary concerns.

IV. CONFUCIAN CRITIQUE ON REGULATIONS FOR HESC RESEARCH IN EAST ASIAN COUNTRIES

According to Professor Shui Chuen Lee's view, a Confucian thinks that we all live in a moral community, which means that we have strong moral commitment to each other; he does not regard society as a group containing atomistic individuals. If some members in our society deeply care about the interests of embryos, we should seriously consider their concerns and not just regard the embryos as biological material but rather as entities that deserve respect and whose interests must also be taken into account:

> Embryos bear a specific relation to us, which greatly increases their moral standing. For embryos bear a much closer relationship to members of our moral community than others and arouse some deep concerns among us ... Furthermore, under such a moral network, greater respect should be given to the moral standing of embryos by our party or other parties. This implies that except when there is no other way possible, we should try to use sources other than embryos to derive stem cells (Lee, 2007).

According to this view, we should avoid deriving stem cells from embryos when other means are available because we must take the interests of those who object to such research into account. Moreover, deriving ES cells from embryos created through SCNT should not be permitted except in a few limited cases because it involves creating a tragedy. While there may be good reason to create this tragedy, it is best to try to explore other options, especially given the objections from the other members of the moral community.

However, as Lee notes, there are still instances when SCNT may be permissible. For example, SCNT may be permissible if it will help to produce a cure for those suffering from genetic diseases.

> On the other hand, creating special embryos from the gametes of those who have genetic diseases for stem cell experimentation becomes a question of justice. It comes close to being a clash between the moral status of embryos and the rights of certain groups of moral agents. It is unreasonable to prohibit their procreation, nor let their progeny carry on the diseases when there are chances for cure. Relieving the pain of others is of paramount value for Confucians. Such pain relieving action carries considerable moral weight and outweighs the moral standing of embryos. The cure of genetic diseases lends support for such research as it may be the best means to end the prevailing tragedies that these ethic groups or families have been facing for generations (Lee, 2007).

Given these considerations, one of the reasons stated by the Bioethics Advisory Committee of Singapore to support the creation of research embryos through SCNT may be morally unacceptable. According to the BAC report, such research can be justified, since therapeutic cloning enables scientists to learn about the mechanisms used to reprogram adult cells to behave like stem cells, potentially making it unnecessary to resort to using embryos as a source of stem cells. However, this reason may be morally objectionable to many of the members of the moral community, because the subject embryos are simply regarded as a means to further ends. Further, this reason does not even claim that such research may contribute to the health of the embryos themselves and their families.

Regarding the guidelines for HESC research provided by the Chinese government, it is unclear under which conditions the creation of embryos for research purposes through IVF or SCNT should be permitted. If all basic or non-clinical research can be permitted to derive ES cells from such embryos, as is the case in Shanghai, which has proposed guidelines supporting the "human-animal" cell fusion technique, then the criteria is too liberal. Such criteria ignore (1) the interests of the subject embryos and their families and (2) the relevant concern to them from members of society. Hence, the Chinese government should further specify the conditions to conduct this kind of research. Moreover, the creation of embryos through SCNT is also not morally acceptable if the reason to do this is simply that ES cells derived from other sources may not be enough for experiments, as is stated in a bill by Congressmen in the Taiwan Parliament.

NOTES

[1]   In Austria, the Reproductive Medicine Act states that cells capable of development may only be used for medical assisted reproduction, but the regulation of HESC research is currently under discussion.
[2]   The Irish Constitution regards the right to life of the unborn as equal as the right to life of the mother, but it is unclear whether the pre-implantation embryo falls into this category, since the term "unborn" has not been judicially interpreted to date, and the Irish Government has appointed the Commission on Assisted Human Reproduction for further examination of this issue.

REFERENCES

Bioethics Advisory Committee (2002). [Online]. Available: bioethics_singapore.org/resources/reports1. html.
Department of Health (2002). 'Ethical Regulations for Embryonic Stem Cell Research,' Taiwan.
Ethical Committee of the Chinese National Human Genome Center at Shanghai (2002). 'Ethical Guide-lines for Human Embryo Stem Cell Research,' China.
European Commission, Directorate General Research (2004). 'Survey on opinions from National Ethics Committees or similar bodies, public debate and national legislation in relation to human embryonic stem cell research and use,' *Volume I: EU Member States*.
House of Lords: The United Kingdom Parliament (2004). 'HFEA grants the first therapeutic cloning licence for research,' *Stem Cell Research – Report*.
Lee, S.C. (2007). *The Family, Medical Decision-Making, and Biotechnology: Critical Reflections on Asian Moral Perspectives*, Springer, Dordrecht.
Legislative Yuan, Republic of China (2003). 'Research and Protection for Human Embryos and Embryoic Stem Cells' [On-line]. Available: *www.ly.gov.tw/index.jp*.
Ministry of Health, Singapore (2003). 'Regulation of Biomedical Research Bill' [On-line]. Available: *www.moh.gov.sg/corp/eservices/econsultations/details.do?id=18*.
Ministry of Science of Technology and Ministry of Health, China (2004). 'Guidelines for Research on Human Embryonic Stem Cells,' *Bioethics Network in China*.
President's Council of Bioethics (2004). United States. 'Monitoring Stem Cell Research' [On-line] Available: http://www.bioethics.gov/reports/stemcell/appendix_c.html.
Walters, L. (2004). 'Human Embryonic Stem Cell Research: An Intercultural Perspective,' *Kennedy Institute of Ethics Journal*, 14 (1), 3–38.

CHAPTER 13

STEM CELL RESEARCH

An Islamic Perspective

SAHIN AKSOY, ABDURRAHMAN ELMALI
AND ANWAR NASIM

Haran University, Turkey and Comstech, Pakistan

Islam is one of today's major religions with more than one billion followers, which constitutes 26% of the world's population. Among every four humans in the world, one of them is a Muslim. The teachings of Islam are quite distinct and thus its ethical standards and views are likely to be different from those of other religions. It is important to emphasize that moral pronouncements and ethical values are intimately related to the beliefs, values and composition of any society. For Muslims there are two major sources of guidance, first is the holy book which gives a complete code of life including economic, social, legal and ethical principles, and second is the Hadith which represents the sayings and the way of life of the holy prophet (PBUH). It is in the light of these above two sources that all ethical dilemmas have to be examined.

There have been numerous major advances in human reproductive biology related to cloning, gene therapy, artificial insemination and surrogate motherhood. A search for answers to these extremely complex questions also leads to other issues such as the patenting of DNA fragments, genetic prescreening for insurance and employment, screening of human populations for early diagnosis and the individual's right to privacy. There are also some specific issues related to human genome project, which merit serious discussion. The present manuscript deals exclusively with the Islamic perspective of the ethical issues related to stem cell research. The key questions that need to be addressed are: What is the moral status of an embryo? When does life begin?

It should be noted that Islamic reflections on bioethics in general and on the use of the human embryo in particular contrast with those of the Confucian authors in this section as they are set within quite different metaphysical and axiological frameworks. On the one hand, Islamic reflections recognize the existence of God and a Divine law that governs all human actions, including medical practice and

S.C. Lee (ed.), The Family, Medical Decision-Making, and Biotechnology, 167–174.

research. On the other hand, Confucian reflections are much more ambiguous about God's existence and the existence of Divine mandates. As a result, the discourse of Islamic bioethics focuses on the Divine law, while the discourse of Confucian bioethics highlights humanity even when it recognizes the mandate of the Heaven.

To understand the Islamic viewpoint we should know the basic principles on which it is based. Islam is an Arabic word (which is composed of two root-words): one Salm, meaning peace and the other Silm, meaning submission. Islam stands for "a commitment to surrender one's will to the will of God: and thus to be at peace with the Creator and with all that has been created by Him. It is through submission to the Will of God which leads to the harmonization of different spheres of life under an all-embracing ideal. Muslims accept the Islamic faith not only as a religion, but also as a way of life.

Islam is a worldview and an outlook on life. It is based on the recognition of the unity of the Creator and of our submission to His will. Everything originates from the One God, and everyone is ultimately responsible to Him. Thus, the unity of the Creator has as its corollary the Oneness of His creation. Distinctions of race, color, caste, wealth and power disappear; our relations with other persons assume total equality by virtue of the common Creator. Henceforth, our mission becomes dedication, worship and obedience to our creator; the Creator becomes our purpose in life.

The norms and assumptions that have characterized belief and action in Islam have two foundational sources. The primary source is the Qur'an revealed by God to the Prophet Muhammad (d.632). The other is the Prophet's actions and precepts, collectively called the *Sunnah*. Muslims regard the Qur'an as the ultimate closure in a series of revelations to humankind from God, and the *Sunnah* as the projection of the life and applications of the Prophet Muhammad.

As we might expect, the principle sources of Islam, namely the Qur'an and *Hadith* (sayings of the Prophet Muhammad (PBUH)) do not refer to the recent development in biotechnology. Therefore, we need to explore other sources, which include *ijma,* the consensus among Islamic scholars of a particular age for the legal ruling applicable to the situation, and *qiyas* known as *ijtihad,* deep and devout reflection from which one can derive an appropriate rule by logical inferences and analogy (Weeramantry, 1988). All lines of Muslim thought converge on a dynamic concept of the universe. Thus Islam rejects a static view of the universe and regards it as always changing and evolving. According to the Qur'an, Change is one of the greatest signs of God. It is explicitly implied in the verse. *"Every day has its own glory".*

Like many other religions, Islam has its own perspective on the status of embryo (Albar, 1992; Duman, 1990; Rispler-Chaim, 1993). Islam considers human life as valuable deserving protection from conception onwards. However, Islam acknowledges some form of dualism, that is, the body and the soul, and their meeting to form a 'full human person'. The creation of human individual and fetal development is referred to in several dozen verses of the Qur'an in various contexts,

(*Al-Sajdah* 32:8–9; *Al-Mu'minun* 23:13–4; *Nuh* 71:14; *Al-Nahl* 16:4; *Al-Qiyamah* 75:37–9; *Al-Tariq* 86:6; *Al-Mursalat* 77:20–1; *Al-Insan* 76:2; *Al-Hajj* 22:5; *Al-'Alaq* 96:1–2; *Ghafir* 40:67; *Al-Zumar* 39:6; *Al-Najm* 53:45–6; *Fatir* 35:11; *Al-i 'Imran* 3:6; *Al-Infitar* 82:68) and in *Hadith*. In particular two verses and two *hadith* are worth mentioning here as they help understand the time of ensoulment and what the status of personhood entails.

The first verse is:

> "*He Who created all things in the best way and He began the creation of man from clay. Then made his progeny from a quintessence of despised liquid. Then He created him in due proportion, and breathed into him of His spirit. And He gave you (the faculties of) hearing and sight and hearts. Little thanks do ye give!*" (Al-Sajdah, 32:7–9)

This verse clearly reports that first the human is "shaped" (in due proportion), then he is "ensouled," and finally the faculties of hearing and sight and hearts are formed. This verse informs us about the "stage" of ensoulment at the intrauterine life.

The second is:

> "*And indeed We created man from a quintessence of clay. Then We placed him as a small quantity of liquid (nutfa) in a safe lodging firmly established. Then We have fashioned the nutfa into something which clings ('alaqa), then We made 'alaqa into a chewed lump of flesh (mudgha) and We made out of that chewed lump of flesh into bones, and clothed the bones with flesh. And then We brought it forth as another creation. So blessed be God, the Best to create*". (Al-Mu'minun, 23:12–4)

This verse is not as open as the first one. It is possible to make different comments regarding the stages of physical developments mentioned in this verse. However, we think it is not that important. The important thing here is the time of these physical developments and we will examine this in depth later.

The role of *Hadith* in Islamic teaching is to help us better understand and interpret the verses of the Qur'an. The Prophet Muhammad (PBUH) was not only a religious or political leader, but also a guide to teach Muslims how to understand Qur'anic verses. Consequently, the Prophet Muhammad (PBUH) is the ultimate interpreter of the Qur'an. He was reported to have said two *hadiths* that we are going to quote. These will later help us to better understand the above Qur'anic verses, especially those concerned with the physical development of the embryo and the time of ensoulment.

The Prophet Muhammad (PBUH) said:

> "*Varily your creation is on this wise. The constituents of one of you are collected for forty days in his mother's womb; it becomes 'alaqa (something that clings) in the same (period) (mithla dhalik), then it becomes mudgha (a chewed lump of flesh) in the same (period) (mithla dhalik). And the angel is sent to him with instructions concerning four things, so the angel writes down his provision (sustenance), his death, his deeds, and whether he will be wretched or fortunate. Then the soul is breathed into him*" (Al-Bukhari; Muslim; Ibn Maja; al-Tirmidhi; Abu Dawud).

And:

> "After nutfa (zygote) has been established in the womb for forty or forty-five nights, the angel comes and says: 'My Lord, will he be wretched or fortunate?' And both these things would be written. Then the angel says: 'My Lord, would he be male or female?' And both these things are written. And his deeds and actions, his death, his livelihood; these are also recorded. Then his document of destiny is rolled and there is no addition to and subtraction from it" (Muslim).

There are different versions (narrations) of both *hadiths* with very minor differences. The first hadith is reported only by Abdullah b. Mes'ud, but the second one is reported by Hudhayfa b. Asid and by some other "companions" (-sahabah- close friends) of the Prophet Muhammad (PBUH). Many Hadith scholars, by referring to the first hadith, and understanding the expression mithla dhalik as 'time equal to this period' rather than 'in the same period', have suggested that the angel comes to the prenate and breath in the soul 120 days after conception. (Rispler-Chaim, 1993) Some earlier scholars (Ibn al-Zamlakani) and contemporary researchers have not agreed on this interpretation (Al-Sawi, 2004; Abdul Rehman, 2004), and concluded from both hadiths that, by understanding the expression mithla dhalik as 'in the same period', the completion of certain physical formation and ensoulment take place after 40 days of conception. As discussed elsewhere to interpret the expression mithla dhaliq as 'in the same period' is more accurate both from an embryological and theological perspective (Aksoy, 2001).

If we accept ensoulment to take place after 120 days, the embryo should look like a nutfa (a drop of liquid; zygote) between day 0 and day 40, it should be something like 'alaqa (something which clings; implantation stage) between day 40 and day 80, and it should be similar to mudgha (a chewed lump of flesh; somites occurrence) in days between days 80 and 120. As we know from the modern embryology these stages occur well before these times (England, 1990). It can be understood from these verses and *hadith* that, in order to receive the soul, i.e., to be a fully human individual person, a prenate must pass the stages of conception, zygote (*nutfa*), implantation ('*alaqa*), somites occurrence (*mudgha*), and beginning ossification and musculation. From the embryological information we have given above, the ensoulment can not take place before 7 weeks after conception, since these embryological stages are not completed before this time (England, 1990).

When the second *hadith* quoted here is examined, it appears that they express very clearly that the angel comes – obviously to give soul – after *nutfa* (zygote) has been established in the womb for 40 or 45 days – or nights. Since the implantation process is completed within nine to ten days of conception, ensoulment takes place sometime between 49–55 days after conception and forms a 'full human person' (Aksoy, 1998). Therefore, according to many Muslim scholars, terminating the life of an embryo before the ensoulment is regarded as disliked (*makruh*), while it is considered as forbidden (*haram*) after this stage (Omran, 1992). This conclusion is particularly important for abortion, cloning and stem cell discussions. Thus, one finds in Islamic bioethics an accent given to the exploration of the Divine law totally absent from Confucian bioethics.

In the light of these discussions it is possible to say that there is no problem to use adult stem cells as long as it does not harm the donor. Islam encourages seeking remedy and treatment as the prophet Muhammad (PBUH) is reported to have said: *"There is a cure for every illness, though we may not know it yet"* (Al-Bukhari). Development of new treatment methods and application thereof, if proven successful, is therefore strongly recommended. Indeed, in the full sense of the Arabic terms used in the relevant Prophetic sayings, the seeking of treatment is commanded rather than merely commended (Abu Dawud). However, there are different views as to the limits on the research for treatments. One is the acceptability or inappropriateness of treatment using forbidden *(haram)* ways. Islamic scholars have referred to the *Hadith* to resolve this issue. In one saying, the Prophet Muhammad (PBUH) is reported to have said: *"God has sent both the disease and cure, and there is a cure for every illness, [therefore] be treated [but] do not to be treated with haram"* (Abu Dawud). Since using adult stem cells for treatment is not principally different than organ transplantation, it is acceptable and advisable from an Islamic perspective.

As far as the usage of fetal stem cells is concerned, the age and the origin of the fetus should be known. If the fetus is the result of a spontaneous abortion (or miscarriage) there is no problem for using it, since it is disposed of anyway. But if it is the result of an induced abortion, then the age of the fetus should be known. If the fetus is younger than 50 days, although it is the consequence of a disliked *(makruh)* action it is lawful to use it. If its age is more than 50 days, since abortion is forbidden *(haram)* after this stage, there is doubt about the appropriateness of its usage. As the *hadith* recommends that one not be treated with forbidden *(haram)* substances, it can be advocated that it is not allowable to use fetal stem cells. But it should be kept in mind that in this case, the forbidden is the "action" not the "substance". Islam considers some substances as "clean" or "pure" *(tahir)* and some of them as "impure", like pigs and their products, and alcohol, due to its nature. The "substance", which is the fetus in this case, is *tahir* (clean; pure). Therefore, provided that the abortion is not initiated for the purpose of obtaining fetal stem cells and that the fetuses will be disposed anyway by using one of the most frequently employed maxims among Islamic legal scholars *(al-darûrât tubîh al-mahzûrât* meaning "necessities render the prohibited permitted")* (Rispler-Chaim, 1993), it is possible to say that the fetal stem cell can be used for therapy. According to one interpreter, "research on stem cells made possible by biotechnical intervention is regarded as an act of faith in the ultimate will of God as the Giver of all life, as long as such an intervention is undertaken with the purpose of improving human health" (Sachedina, 2000).

In the case of using embryonic stem cells, we need to know how the embryos are produced. Since the embryos used in embryonic stem cell research are mostly spare embryos from IVF treatments, it follows that in the Islamic approach to IVF treatment, such practice is legitimate as long as it is performed between a living husband and wife in mutual agreement. However, producing spare embryos and then storing them in a freezer for future uses is not acceptable (Sachedina, 2000, pp. 19–27). According to the Islamic Fiqh Association (IFA) in Jeddah in association with the Medical Fiqh meeting in Kuwait in the year 2000, if there are excess embryos in any shape or form, they will be left without medical intervention to end their life naturally (Islam Online). However it is not possible to accept this view since this argument is based on the fear of the "misuse" of embryos rather than a theological or philosophical reasoning. It is apparent from the argument that allowing the embryo to die on its own is not refuted on the stance that it is killing a potential human being. Therefore, it is hard to see what the difference is, although some may argue that the use of embryonic stem cells is the killing of unused embryo by means of taking the cells. However, allowing the embryo to die is also done by the means of not taking any steps to save the potential human being.

To better understand our objection to the decision of IFA, it is of benefit to bear in mind that expert opinion is always well considered in the Islamic tradition. When an expert (i.e., in this case a physician) talks in a conscientious manner about a particular subject, the people of the religious community usually listen to what he has to say (Aksoy & Elmali, 2002). Therefore, if a responsible physician suggests that to create spare embryos is necessary for the success of IVF treatment, then this suggestion can be followed. However, it is safe to say that doctors should create the minimum amount of the required number of embryos to avoid unnecessary wasting of human lives. In addition, the couple should also be informed that, if everything goes well, they are consenting to all the embryos to be implanted. If despite all these precautions there are spare embryos that will be destroyed, these embryos can be used in stem cell research, since they have not been ensouled yet and are therefore not full human persons. This suggestion was also supported by Basalamah, who argued that "At the point when a fertilized egg has reached stage 8 and has divided into 32 cells only, no limbs or organs are yet formed. Therefore, at least at present, it is not possible to use surplus eggs for organ transplant, although it is possible to use some of their cells. Transplanting such cells is far better than destroying them, which would be comparable to infanticide" (Sheikh Muhammad al-Mokhtar al-Salami Mufti).

This is an extremely important point since this clearly relates to the ultimate objective or purpose of any undertaking. Islamic teachings place a great deal of emphasis on "Neeat – Intention" Allah's Apostle said, *"The reward of deeds depends upon the intentions and every person will get the reward according to what he has intended".* Production of stem cells at a commercial level is an industrial setup with the purpose of exploiting others for one's own commercial and monetary benefits; this is obviously something that Islam will not approve. However, limited and defined supplies for curing serious diseases will be acceptable. In any such situation a careful risk/benefit analysis becomes essential, including an analysis of the motive. This is only one of the major considerations that need to be brought into focus while examining the moral or ethical questions that relate to human reproductive biology. An in-depth ongoing critical analysis and continuing dialogue between religious scholars and researchers is required to develop a consensus and to come up with specific answers. In Islam, the concept of *Ijtehad* provides an opportunity and flexibility to achieve these worthwhile goals.

To conclude, science and technology continue to push the boundaries to pursue its goal, which is to know the unknowable and to do the undoable. In Islam, it is believed that man cannot do the things that are intrinsically undoable, and cannot know the things that are intrinsically unknowable. Therefore, if man can do and know a thing, then it is nothing but doable and knowable in its essence. As we have seen, Islamic reflections on the allowable use of the human embryo is thus set in a moral analysis that has at its basis recognition of Divine law that is different but comparable with the Confucian reflections on the mandate of Heaven. According to Islam, everything on earth is created by God for the service of mankind; provided that they are not exploited. It is thought that science is the business to explore the secrets and laws of nature, which were set by God. Islam always encourages man to contemplate and explore the new horizons. Stem cell research is one of the new horizons, and Islam does not object to it in the first encounter. All kinds of actions are principally permissible in Islam as long as it is not categorically prohibited. Therefore, stem cell experimentation is allowable from a perspective of Islamic teaching as long as it is with the purpose of improving human health. However it still needs to be examined over and over again, and after that a final word can be said.

REFERENCES

Authors Note: Not all publication names are listed for the books referenced. These books are mostly Arabic and do not have names of the publishers and/or places of publication. However, these are quite common in Islamic literature and do not prevent people from reaching these references.

Abdul Rehman, O. (2004). *Does the Qur'an Plagiarize Ancient Greek Embryology?* [On-line]. Available: http://www.aquaire.clara.co.uk/

Abu Dawud, *al-Sunan, Kitab al-Sunnah.*

Aksoy, S. (2001). 'A Critical Approach to the Current Understanding of Islamic Scholars on Using Cadaver Organs without Prior Permission,' *Bioethics*, 15:5/6, 461–472.

Aksoy, S. (1998). 'Can Islamic Texts Help to Resolve the Problem of the Moral Status of the Prenate,' *Eubios Journal of Asian and International Bioethics*, 8:3, 76–79.

Aksoy, S. & Elmali, A. (2002). 'Four Principles of Bioethics' as found in Islamic Tradition, *Medicine and Law*, 21, 211–224.

Albar, M. (1992). Human Development, As Revealed in the Holy Qur'an and Hadith, Saudi Publishing House, Jeddah: 57–62.

Al-Bukhari, *al-Sahih, Kitab al-Tib.*

Al-Sawi, Abd al-jawad (2004). 'Atwar al-janin wa nafkh al-ruh' (Development Stages of Prenate and Ensoulment.) [On-line]. Available: http://www.islamtoday.net/articles/show_articles_content.cfm?artid=942&catid=74 (in Arabic). Accessed May 25, 2004.

Duman, M. (1990). *Kur'an-i Kerim ve Tibba Gore Insanin Yaratilisi ve Tup Bebek Hadisesi* (The Creation of Man and Tube Babies According to Qur'an), Nil Yayinlari, Izmir: 8–20 (in Turkish).

England, M. (1990). *A Colour Atlas of Life before Birth*, Wolfe Medical Publications, London.

Holy Qur'an; Al-Sajdah 32:8–9, Al-Mu'minun 23:13–4, Nuh 71:14, Al-Nahl 16:4, Al-Qiyamah 75:37–9, Al-Tariq 86:6, Al-Mursalat 77:20–1, Al-Insan 76:2, Al-Hajj 22:5, Al-'Alaq 96:1–2, Ghafir 40:67, Al-Zumar 39:6, Al-Najm 53:45–6, Fatir 35:11, Al-i 'Imran 3:6, Al-Infitar 82:68.

Islam Online. Available: http://www.islam-online.net/fatwaapplication/arabic/display.asp? hFatwaID= 14717. (in Arabic) Accessed: January 25, 2003.

Ibn Maja, *al-Sunan, Kitab al-Muqaddima.*

Ibn al-Zamlakani, *al-Burhan al-Kashif an I'jaz al-Qur'an* (Discovering Evidence for the Inimitability of the Qur'an):275 (in Arabic).

Muslim, *al-Sahih, Kitab al-Qadar.*

Omran, A. (1992). *Family Planning in the Legacy of Islam* (pp.190–193), London, Routledge.

Rispler-Chaim, V. (1993). *Islamic Medical Ethics in the Twentieth Century*, E.J. Brill, Leiden.

Sachedina, A. (2000). 'Islamic perspectives on research with human embryonic stem cells,' in National Bioethics Advisory Commission, *Ethical Issues in Human Stem Cell Research. Vol. III. Religious Perspectives*, Rockville, MD, US Government Printing Office: G1–G6.

Sheikh Muhammad al-Mokhtar al-Salami Mufti of the Republic of Tunisia. 'The Transplantation of Nerve Cells, with Particular Reference to Brain Cells' [On-line]. Available: http://www.islamset.com/bioethics/organ/salami.html#2. Accessed: January 22, 2003

Weeramantry, C. (1988). *Islamic Jurisprudence: An International Perspective*, Macmillan Press, London, Chapter 3.

CHAPTER 14

WHY WESTERN CULTURE, UNLIKE CONFUCIAN CULTURE, IS SO CONCERNED ABOUT EMBRYONIC STEM CELL RESEARCH
The Christian Roots of the Difference

H. TRISTRAM ENGELHARDT, JR.

I. WHY THE WEST IS SO DIFFERENT

Francis Fukuyama notes that much of the moral and bioethical discussions in the Pacific Rim has a character noticeably different from that of the West. "A number of countries in Asia...for historical and cultural reasons have not been nearly as concerned with the ethical dimensions of biotechnology...[This culture] lacks religion per se as it is understood in the West – that is, as a system of revealed belief that originates from a transcendental deity" (Fukuyama, 2002, p. 192). The essay by Aksoy, Elmali, and Nasim in this volume shows one possible source of this contrast. The Abrahamic religions traditionally appreciate morality as rooted in divine commands given by a personal God to man. This moral-metaphysical insight still shapes the West. The Islamic appreciation of sharia is a variation on this understanding at the root of Judaism and Christianity. The Confucian cultural tradition does not recognize the transcendent personal God of the Abrahamic religions, nor His commandments.

Over against the Athenian and later Roman aspiration to a morality grounded in a common, discursively justifiable, immanent rationality, the traditional Christian appreciation of morality is grounded in an experience of a transcendent God. This experience recognizes God as the transcendent Creator Who is the source of all that is immanently good and right. The result is that, *pace* Plato's Euthyphro, the good, the right, and the virtuous are not simply such because God wills them. Nor does God will the good, the right, and the virtuous simply because they are such. Rather, since that which is created can only be fully and rightly appreciated in terms of the Creator, the good, the right, and the virtuous can only be fully and rightly appreciated in terms of the holy. Created being is one-sidedly and incompletely understood apart from a consideration of its Creator. Moreover, the God to Whom all created being is in the end directed is personal. The Truth is thus ultimately a

S.C. Lee (ed.), The Family, Medical Decision-Making, and Biotechnology, 175–181.

Who, not a what (Sophrony, 1988). Integral to this metaphysics and axiology is a "mystical" theology, a recognition of the capacity of humans to have noetic, that is, non-sensible, albeit empirical knowledge (Hierotheos, 1998).

In contrast to this moral-metaphysical appreciation, there is an account of moral claims grounded in the synthesis of Platonic, Aristotelian, and Stoic that was integrated with Christian commitments, thus producing Roman Catholicism, along with its views regarding natural law. The results of this cultural-religious synthesis begin to take a distinctive form at the beginning of the second millennium. In time, these commitments shaped the Western European expectation that one can ground a canonical, secular morality through sound rational argument, without an experience of God. This moral understanding rooted in the Academy, the Lyceum, the Stoa, and the Roman Catholic medieval, moral-metaphysical synthesis overlays remnants in the West of an older, competing moral understanding rooted in the Christianity of the last millennium. This later understanding, grounded in an experience of the transcendent God, locates all appropriately oriented accounts of proper behavior in terms of an encounter with the Holy, Who is the God Who commands. This divine-command orientation of morality places bioethics with reference to a set of covenants ranging from that given to Adam to those given to Noah and Moses, and then to all men through Christ. This appreciation of morality and covenant, which is lodged in a noetic theological experience, though obscured in Western Christianity by the early second-millennium philosophical-theological synthesis of the Western Middle Ages, still sustains a moral perspective that contrasts with that of Confucian understandings.

In terms of a relationship with a God Who commands, one can appreciate the warnings of St. Basil the Great in Letter 188 that abortion, indeed all killing of early human embryonic life, is to be treated as murder, even if the embryo is not ensouled. At issue is not just an affront against human life, but an affront against the ways in which humans are to live towards God.[1] There is an appreciation that such moral prohibitions recognized what is involved in rightly aiming at holiness. Finite humans are unable fully, non-onesidedly, and non-incompletely to appreciate the good, the right, and the virtuous, save with reference to the transcendent Creator-God, Who as transcendent escapes their attempts systematically to appreciate the good, the right, and the virtuous. Given the background-framing theology of the Christianity of the first centuries, there remains even in the secular West a persistent strand of appreciation, however marginalized, of the moral significance of taking early human embryonic life. This appreciation reverberates as a cultural recollection from Europe's Christian past, as a moral intuition of the evil of destroying early human embryonic life. Because this intention is no longer located in its appropriate, sustaining, epistemological, and metaphysical framework, it is often both distorted and obscured. Impelled by the influence of the Western philosophical-theological medieval synthesis, attempts are made to frame general philosophical supports for this intuition. Yet, since these attempts lack the appropriate supporting epistemo-logical and metaphysical framework, they can only be partially successful. There is a recognition that there is something of moral significance at stake, but no longer an appreciation of its full character.

Western concerns regarding early human embryonic life and abortion provide a complex contrast with the predominant morality of the Pacific Rim. In particular, China's Confucian heritage does not directly thematize the evil of taking early human life. In addition, there is no explicit recognition of the personal God Who commands obedience (though some recognition of His presence persists).[2] Nor is there a natural-law morality as this developed in the Western European Middle Ages and persisted in Roman Catholic theology. As a result, health care policy reflections on issues such as human embryonic stem cell research proceed largely without significant controversy, or at least without the level and character of controversy that mark Western culture. This is the case, though the presence of Islam in the Pacific Rim and its recognition of God as the source of moral commands, of sharia, as the article by Aksoy, Elmali, and Nasim demonstrates, provides an additional moral framework robustly different from that of the predominant Confucian moral-philosophical discourse. Over against Confucian moral and metaphysical commitments, there is Islam, Judaism, and Christianity. Moreover, there is not just one Christianity, but the Christianity of Orthodox Christianity, which reflects the Christianity of the first millennium over against the multiple and diverse Christianities of the West.

II. JERUSALEM, ATHENS, AND QOFU

The full texture of Judeo-Christian prohibitions concerning abortion and the destruction of early human life is more complex than is often appreciated. In part, this complexity is obscured by incomplete portrayals of Jewish views regarding the taking of human embryonic life that fail to attend to basic distinctions grounded in specific covenantal requirements.[3] Many reflections have relied on what Jews understand to be prohibited and allowed for Jews, that is, what follows from halakhah[4] bearing on Jewish conduct, but without attention, or only meager attention, to halakhah bearing on the conduct of Gentiles. This failure has tended to obscure the circumstance that Orthodox Jewish reflections, unlike the philosophical tradition that grew out of Greek intellectual concerns centered in Athens, and which later framed Roman Catholicism and thus much of Western Europe, does not aspire to a primarily discursively grounded, immanent set of prohibitions and moral injunctions. The focus is on a set of demands made by a personal God, not on a set of principles purportedly fully vindicated by sound rational argument. God is recognized as fully transcendent and not an object to be grasped adequately by natural theology.

Because of the covenantal character of morality and the transcendent, personal character of God, Orthodox Judaism has neither a morality nor a theology, as one would understand these within the West. The difference lies in great measure in the latter's commitment to a moral philosophy and a philosophical theology. Instead, Orthodox Judaism recognizes moral restraints as grounded in laws given by God either to Noah or to Moses, as does traditional Christianity, which recognizes the final covenant given by Christ. The result is that Orthodox Jews appreciate that the prohibitions and injunctions binding Gentiles (i.e., bnai-Noah are bound to obey the seven laws given to them) are on many points different from those binding Jews (i.e., Jews are obliged to keep the 613 laws given to Moses). Orthodox Jews do not affirm one universal code of behavior binding all humans. Though moral theorists might find the law of Moses and its halakhah heuristic for general moral reflections, that law does not determine the specific bioethical obligations binding Gentiles (Brody, 1983). Given these background distinctions, from the perspective of Orthodox Judaism one confronts at least three genre of moralities along with their bioethics: (1) prohibitions and injunctions binding Jews grounded in the law given to Moses; (2) prohibitions and injunctions binding Gentiles (bnai-Noah) grounded in the law given to Noah; and (3) alleged prohibitions and injunctions grounded in particular, moral-philosophical understandings such as those which grew out of Athens and Rome (and which have to some extent come to exist within certain strands of contemporary Confucian thought).

With respect to Jewish views about the morality of abortion as well as actions against early human embryonic life, the Halakhah binding Gentiles are quite stringent, even though the Halakhah binding Jews have over time come to be somewhat indulgent.[5] The result is that there are two parallel religious sets of obligations with regard to early embryonic life, grounded in two different covenants or sets of commands, depending on whether one's obligations are those of a Jew or those of a ben-Noah (Brody, 1983). Most significantly, as Brody points out, the Talmud appreciates that the destruction by bnai-Noah of a human embryo counts as murder. "On the authority of R. Ishmael it was said: [He {a ben-Noah} is executed] even for the murder of an embryo. What is R. Ishmael's reason? –Because it is written, *Whoso sheddeth the blood of man within [another] man, shall his blood be shed.* What is a man within another man? –An embryo in his mother's womb" (*Sanhedrin* 57[b]). In this light, one can understand the Christian covenant and the obligations that it imposes on humans to God regarding early embryonic life as realizing for all the prohibitions that bound bnai-Noah.

The framework within which Orthodox Jewish and traditional Christian moral concerns are lodged thus contrasts epistemologically and metaphysically with that born of Athens and of Greco-Roman philosophical reflections generally: though the latter recognized the existence of God, God was largely acknowledged only as a philosophical principle and surely not as the personal lawgiver who can be encountered person to person. The focus was on answering moral questions in terms of the relationship among a set of considerations and reasons rather than in terms of the relationship of finite persons with a transcendent, personal God. The contemporary, secular moral philosophy of the West in a number of steps transformed this Greco-Roman tradition, so that by the beginning of the 20th century references to God in professional philosophical reflections became increasingly infrequent. Morality came to be framed in terms of a practical agnosticism, so that there was an attempt to speak of the good, the right, and the virtuous as if reality had no final meaning, as if everything ultimately came from nowhere, goes nowhere, and for no final purpose. One is enjoined to act as if God did not exist. There appear to be at least some grounds to hold that this view was not that embraced by the preponderance of the original Confucian tradition.[6] A major task for those wishing to gauge the force of the Confucian cultural inheritance will be to reconstruct the authentic and integral character of the views of Confucius and their implications for a proper appreciation of the good, the right, and the virtuous (Fan, 2007a).

III. A CONCLUDING PUZZLE: IDENTIFYING THE MORAL RESOURCES OF CONFUCIANISM

A significant difficulty confronts those assessing Confucianism's resources for appreciating the moral status of human embryonic life: academic reflections on Confucian thought have in great measure attempted to accommodate to the conceits of the contemporary secular West.[7] In part, Confucian reflections have passed through a reformulation with similarities to the philosophical-theological synthesis of the Western Middle Ages and the European Enlightenment: they have been recast by later influences within Chinese thought, as well as by philosophical and cultural aspirations imported from the West. European cultural commitments from Athens have also been joined with later Chinese thought to reshape Confucian appreciations of morality and metaphysics.[8] The result is a widespread attempt to recast Confucian thought in the image and likeness of European social-democratic theory, so that Confucian reflections will not appear out of place when compared to the contemporary conceits of the West. Much of Confucian reflection has been shaped by a powerful and largely successful Euro-centrism.

These developments to the contrary notwithstanding, as Ruiping Fan indicates in his reconstructionist Confucian approach, authentic Confucianism is not as compatible with secular European thought as many later strands of the Confucian tradition might suggest (Fan, 2007a). In particular, the contrast between Western cultural concerns for early embryonic life and for the requirements of God on the one hand, and those of Confucian thought on the other, may not be as stark as this might at first blush appear. Though it is the case that "it is impossible for Confucianism to have an explicit prohibition on abortion from a personal God as in the case of Judeo-Christianity" (Fan, 2007b, p. 140), on the other hand, "it is not that Confucianism does not have capacity to appreciate the moral significance of early human embryonic life...In this regard the contrast between the Abrahamic religious concerns with early embryonic life and the Confucian thought on the issue may not be as stark as it might appear at first blush. It might have been an entire misinterpretation that Confucianism does not take early embryonic life seriously" (Fan, 2007b, p. 140, 144). As Fan puts the matter, "All these considera- tions indicate that it would be very difficult to construct a convincing double-effect argument from the Confucian moral resources to support the killing of embryos in this research case [the creation of human embryos for embryo stem cell research]" (Fan, 2007b, p. 143).

This is a matter to be further explored by Confucian scholarship as it assesses the significance of such key issues as the family, virtue, and the dictates of Heaven, as well as the phenomenology of Confucianism's appreciation of the moral significance of early human embryonic life. A reassessment of the metaphysical commitments of Confucian thought and of its ability to recognize the moral significance of early human embryonic life will contribute significantly to contemporary moral theory and political philosophy. No matter how this reassessment of Confucianism's recent accommodations to contemporary Western secular philosophical fashions proceeds, there remains a substantive contrast between the metaphysics and axiology lying at the roots of moral discussions within the ambit of the Confucian cultural heritage, versus that within the cultural heritage of the Abrahamic religions. The first is not embedded within the framework of a culture that has given central regard to the transcendent, personal God Who commands. The second is. This circumstance, as Fukuyama correctly observes, lies at the roots of major moral cross-cultural differences.

NOTES

[1] The evil of sin is first and foremost that it aims us away from God. Thus, King David, a murderer and adulterer, can say, "Against Thee only have I sinned and done this evil in Thy sight" (Psalm 50, LXX).

[2] The extent to which Confucianism is committed to the recognition of the existence of God or at least of a personally responsive Heaven is a complex issue. For one view on this matter, see Louden, 2002. See also Ching, 1977; Fung, 1983; Ivanhoe, 2006; and Legge, 1971.

[3] David Feldman, for example, in his treatment of abortion does not explore in any detail the stark difference between halakhah for Jews and those for bnai-Noah (Feldman, 1986, pp. 79–90). He does in passing, note the difference between these two laws and their implications for appropriate human behavior (Feldman, 1986, p. 59).

[4] The term halakhah refers to the law of Moses and the law given to Noah as well as the rabbinic rulings made regarding their application.

[5] For the halakhah bearing on Jews regarding abortion, see Jakobovits, 1962, pp. 182–191.

[6] There is evidence that, though some later Confucian scholars acted in disjunction from any belief in God or in the importance of Heaven, this was not the position taken by Confucius himself. For example, Fan states that "the existence of the family reflects a profound moral structure and significance set by Heaven (tian)" (Fan, 2007b, p. 131).

[7] Among many scholars and contemporary Confucians, there has been an attempt to construe Confucian thought as able to support contemporary Western moral and philosophical fashions, rather than as bringing a set of insights that can critically recast those fashions. See, for example, Bell, 2000; Hall and Ames, (1999); and Wang, 2003.

[8] For an example of an attempt to accommodate Confucian thought to Western European expectations, one might think of Liu, 1999.

REFERENCES

Bell, D.A. (2000). *East Meets West*, Princeton University Press, Princeton, New Jersey.

Brody, B.A. (1983). 'The Use of Halakhic Material in Discussions of Medical Ethics,' *Journal of Medicine and Philosophy*, 8, 317–328.

Ching, J. (1977). *Confucianism and Christianity*, Kodansha International, New York.

Confucius (1971). 'Prolegomena,' in J. Legge (trans.), *The Analects of Confucius: Confucian Analects, the Great Learning & the Doctrine of the Mean* (pp. 97–101), Dover Publications, New York.

Epstein, J. (Ed.) (1987). *Tractate Sanhedrin*, J. Shachter & H. Freedman (trans.), Soncino Press.

Fan, R. (2007a). *Reconstructionist Confucianism: The Rebirth of an Ancient Ethics*, Springer, New York.

Fan, R. (2007b). 'The Ethics of Human Embryonic Stem Cell Research and the Interests of the Family,' in S. Lee (Ed.), *The Family, Medical Decision-Making, and Biotechnology*, Springer, New York.

Feldman, D.Z. (1986). *Health and Medicine in the Jewish Tradition*, Crossroad, New York.

Fukuyama, F. (1993). *The End of History and the Last Man*, Harper Perennial, New York.

Fukuyama, F. (2002). *Our Posthuman Future*. Farrar, Straus and Giroux, New York.

Fung, Y. (1983). *A History of Chinese Philosophy*, Derk Bodde (trans.), Princeton University Press, Princeton, NJ.

Hall, D.L. & Ames, R.T. (1999). *The Democracy of the Dead: Dewey, Confucius, and the Hope for Democracy in China*, Open Court, Chicago.

Hierotheos [Vlachos]. (1998). 'The Mind of the Orthodox Church,' in E. Williams (trans.), *Birth of the Theotokos Monastery*, Levadia, Greece.

Ivanhoe, P. (2006). 'Heaven as a Source for Ethical Warrant in Early Confucianism,' *Dao: A Journal of Comparative Philosophy*, 6.1.

Jakobovits, I. (1962). *Jewish Medical Ethics*, Block, New York.

Liu, S. (1996). 'Confucian Ideals and the Real World,' in W. Tu (Ed.), *Confucian Traditions in East Asian Modernity* (pp. 92–111), Harvard University Press, Cambridge, Mass.

Louden, R.B. (2002). '"What Does Heaven Say?": Christian Wolff and Western Interpretations of Confucian Ethics,' in B.W. Van Norden (Ed.), *Confucius and the Analects* (pp. 73–93), B. Oxford University Press, New York.

Sophrony [Sakharov]. (1988). *We Shall See Him as He I*, in R. Edmonds (trans.), *Stavropegic Monastery of St. John the Baptist*, Essex.

Wang, J. (2002). 'Confucian Democrats in Chinese History,' in D.A. Bell and C. Hahm (Eds.), *Confucianism for the Modern World* (pp. 69–89), Cambridge University Press, New York.

CHAPTER 15

CONFUCIAN HEALTHCARE SYSTEM IN SINGAPORE

A Family-Oriented Approach to Financial Sustainability

KRIS SU HUI TEO

City University of Hong Kong, Hong Kong, PRC

I. INTRODUCTION

Healthcare issues form the core of any government's agenda, and a good government is expected to take care of its people and keep them healthy. However, the burden of healthcare is a much contested point and politicians remain divided on the amount the people have to pay for healthcare out of their own pockets. On the one hand, the government must allocate healthcare resources to ensure that each person gets an adequate and appropriate amount of healthcare. On the other hand, measures must be taken to prevent an over reliance on the government and possible abuse of healthcare facilities. In places where the population is aging at a fast pace, effective healthcare policies become even more pressing and governments have to come up with manageable solutions in the long-run that are both of high quality and cost effective.

There are three main parties involved in the determination of healthcare responsibility – the government, the people and the healthcare provider. Some common healthcare schemes are mandatory healthcare funds, healthcare insurance and welfare schemes. Governments generally choose a combination of any of the above schemes.

In this paper, the healthcare systems of Singapore and Hong Kong are chosen as cases for comparison because both places have similar historical and geographical backgrounds but very different policies on healthcare. While the Hong Kong government provides highly subsidized or free[1] healthcare to its people, the Singapore government has adopted a mandatory healthcare savings plan for individuals, transferable among immediate family members. This paper shall first attempt to exposit the Hong Kong and Singapore healthcare systems by tracing their historical backgrounds that have shaped current healthcare policies. The Hong Kong and Singapore healthcare systems then will be compared from an

183

S.C. Lee (ed.), The Family, Medical Decision-Making, and Biotechnology, 183–195.
© 2007 *Springer.*

economic perspective on issues such as family savings, cost containment, sustainability, government subsidies and the private sector. It shall be argued that Hong Kong's stance on highly subsidized healthcare is not sustainable in the long run as recent incidents have demonstrated, culminating in calls for its reform from various spectrums of society. The feasibility of the Singaporean system of healthcare, which relies primarily on family savings, is considered to be a better system, economically. Finally, the moral aspect of the Singaporean healthcare system shall be explored vis-à-vis Confucian ethics. In my opinion, Singapore's healthcare system is both economically feasible and morally sound and offers a good example for other countries to learn from, especially if they are combating critical healthcare problems such as rising expenditures and an aging population.

For better clarity and focus, I shall be focusing on inpatient instead of outpatient healthcare because the former often requires more resources in terms of facilities, manpower, expertise, financing, etc. As a result, inpatient healthcare is the area where the government has to come in on a large scale of operation and consequently, inpatient health care is of immense significance to policy formulation. The public sector provides about 80% of hospital care in Singapore while in Hong Kong, the percentage stands at a high 95%. In terms of primary healthcare, public providers account for 20% of the total share in Singapore and 43.6% in Hong Kong (Ministry of Health, Singapore, 2005; Medical Development Advisory Committee, 2005).

II. HONG KONG HEALTHCARE SYSTEM

The government of Hong Kong did not actually take a proactive approach in the formulation and implementation of healthcare policies; neither did the colonial bureaucrats lay down strict guidelines for the formulation of healthcare policies. In the 1960s, the Hong Kong government finally announced that its mission was to provide "low cost or free medical and personal health services to that large section of the community which is unable to seek medical attention from other [private] sources" (Gauld, 1997, p. 26).

Since then, Hong Kong healthcare has been operating as a broad welfare system, with the government providing highly subsidized healthcare services to its residents. The Hospital Authority (HA) was set up in 1990–91 as an independent body to effectively coordinate and manage public hospitals. The HA management board is accountable to the government. Public hospitals account for about 95% of all inpatient bed days in Hong Kong and 2003 figures show that the government subsidized about 97% of the total cost of inpatient care, arguably one of the highest rates of subsidy in the world (Health and Medical Development Advisory Committee, 2005).

In recent years, especially after the Asian Financial Crisis in 1997, which consumed much of the government's reserves and caused huge budget deficits, healthcare issues have become very pressing in Hong Kong. In addition, burgeoning costs due to more advanced medical trainings, treatments and facilities, and an aging population also lead to the Hong Kong government's need for various reforms of its healthcare system in order to achieve long-term stability. "From a user's perspective, the greatest problem is quality of service ... Users of public services complain of long waiting times, an indifferent service attitude and lack of choice, while those who go to the private sector face high costs and variable service quality" (Wong, 1999).

Numerous measures are sought to reduce direct government expenditure on healthcare. One of the first measures to be implemented was an increase in the amount a Hong Kong resident has to pay for emergency healthcare services. This created quite a stir among Hong Kong residents who are used to enjoying free emergency healthcare services. Fee structural changes to prescriptive drugs in public hospitals are also being proposed and implemented, but the changes are met with equal resistance.

The Mandatory Provident Fund (MPF) came into operation on December 1, 2000, to provide a more reliable form of retirement protection across the board. It is targeted at working people aged between 18 and 65, except for those whose salary level fall above HKD 20,000 or below HKD 5,000.[2] Each month, 5% of their salary will be deducted and the employer will contribute a matching amount to be deposited in the MPF accounts. Upon retirement, members can withdraw their MPF in one lump sum to be used at their own discretion. MPF is a privately-owned investment fund and members can choose between various investment options such as the Capital Preservation Fund, the Money Market Fund, the Guaranteed Fund, the Bond Fund, the Equity Fund, etc. Although there is no separate comprehensive healthcare plan under MPF, it is also intended to help with the increased healthcare needs of a retired person, thereby lessening the healthcare burden on the government.

III. SINGAPORE HEATLHCARE SYSTEM

When Singapore's Minister Mentor Lee Kuan Yew was a student in Cambridge, he had a first encounter with Britain's National Health Service and was deeply impressed: "Soon after the National Health Service Act was passed in 1948, I went to collect my spectacles from an optician in Regent Street ... the optician proudly told me that I did not have to pay for them ... I was delighted and thought to myself that this was what a civilized society should be" (Lee, 1998, p. 129).

Idealism aside, Mr. Lee soon realized that Singapore could not possibly implement an across the board, welfare-based healthcare system over the long run due to a number of reasons. Singapore was a young nation and tax incentives must be given to attract workers to the country. The government was working with a tight budget and the high population density further strained healthcare resources. While the government did much to improve the overall health conditions such as educating the public, increasing health awareness, good sanitation and infrastructure, the people also have to be encouraged to take responsibility for their family's health.

Singapore is fortunate in that as a young country, it can learn from the experiences of many developed and developing countries in healthcare strategies. The Singapore government has learnt to take proactive approaches to preempt rising government expenditures on healthcare. The government of Singapore aspires to maintain a proactive stance on healthcare, and one of its primary concerns is for prevention and health education to be taught in schools to instill proper health knowledge in the young.[3] Where policy is concerned, the government tries to ensure that one has enough savings to take care of one's health and the health of one's family.

In a 2001 study done by Canadian health economist, Cynthia Ramsay, Singapore's healthcare system was ranked first when compared to Canada, the United States, the United Kingdom, Switzerland, Germany, Australia and South Africa. This has debunked the myth that more spending on healthcare is better than less, due to the many benefits of containing healthcare expenditures. Singapore's way of placing responsibility on patients to finance at least a portion of the medical costs also proved to be favorable.

Singapore has a three tiered healthcare system. The patients or their families bear the major responsibility, making up the patient-based financing. When it comes to non-patient based financing, the private sector employer provides for the healthcare of its employees mainly through Central Provident Fund contributions and corporate insurance schemes. A major provider of employment insurance is the National Trade Union Corporation (NTUC) through NTUC Income, an insurance company founded on co-operative principles. Finally, the government provides healthcare subsidies at 20% to 80% for all classes of ward, except Class A which does not receive any subsidy,[4] and offers a safety net for healthcare so that no one is denied treatment. More importantly, the government has also come up with policies to help individuals save up for major financial needs. One main policy is the Central Provident Fund.

The Central Provident Fund (CPF), which was set up in 1955, is a mandatory social security savings scheme to which both employers and employees contribute. The CPF can be used to pay for public home-ownership, medical and term-life insurance, limited investments, local education and hospitalization. CPF savings generate interest at market-related rates for their members annually. Currently, an employee contributes around 20% of her income to the CPF account each month and the employer contributes around 13% of her salary to the account.[5] Of interest to our present discussion on healthcare, CPF medical insurance (Medishield), hospitalization scheme (Medisave) and elderly care (Eldershield) will be elaborated.

According to a pioneering 1994 report by the World Bank on how to avert problems associated with aging population, the twin criteria of protecting the old and promoting economic growth are recommended. In addition, it was proposed that governments provide comprehensive retirement protection by using the "three-pillar" approach of a publicly managed, tax-financed social safety net for the old, mandatory, privately managed, fully funded contribution scheme, and voluntary personal savings and insurance.

Extending the "three-pillar" approach specifically in the area of healthcare provision, the Singapore government has come up with a comprehensive framework through Medisave, Medishield, Eldershield and Medifund. All the schemes are run through individuals' CPF savings, except for Medifund which the government makes contributions to.

The Medisave scheme, introduced in April 1984, is the main component of the healthcare savings scheme implemented on a nation-wide scale. The monthly contribution to the Medisave account is tied to the CPF contribution and it stands at around 6 to 8% of one's salary.

Medishield is a low-cost insurance scheme which covers one's hospitalization costs. It was introduced in July 1990 to provide CPF members with affordable insurance plans against catastrophic illnesses where the Medisave accounts would not be sufficient to cover hospital expenses. Basic Medishield coverage is enough for Class B2/C ward while Medishield Plus coverage is offered at a higher premium for those who prefer services at Class A/B1 level.[6]

Eldershield was introduced in June 2002 to automatically cover all Singaporeans and Permanent Residents reaching 40 years of age and who have CPF accounts against severe disability or illness due to old age. Like Medishield, it is an opt-out scheme. Eldershield helps by giving monthly cash handouts and taking care of out-of-pocket expenses. It is a basic insurance plan provided by two companies (Great Eastern Life and NTUC Income) with affordable premiums deductible from the Medisave account.

The basic healthcare safety net comes in the form of Medifund, an endowment fund set up by the Singapore government in 1993. It is intended as a last resort to help people who are unable to pay for their own medical expenses despite government subsidies, Medisave and Medishield. Eligibility for Medifund is assessed by medical social workers.

Having gone through the healthcare systems of Hong Kong and Singapore, comparisons will be made on the two systems in the following analysis.

IV. HONG KONG AND SINGAPORE HEALTHCARE SYSTEM COMPARED

4.1 Role of Family

Mr. Lee Hsien Loong, in his first Prime Minister Chinese New Year message, asserted the importance of the family in Singapore and emphasized that any assistance given by the government "can only complement the family's traditional role, and not supplant it" (Lee, 2005, p. 1). Singapore encourages its people to take greater responsibility for both their own and their families' health. The Medisave scheme is primarily a family savings healthcare scheme that extends to immediate family members. In the event of any member of the family falling ill, the spouse, children, parents and even grandparents can contribute to the costs of incurred healthcare. Under the circumstance that the combined savings from this group of immediate family members are not enough to meet the costs of healthcare, the Medisave accounts of other close relatives such as sisters and brothers can also be used.

The people of Hong Kong, on the other hand, generally rely heavily on the government for healthcare and family savings are normally used for other purposes such as education, mortgages, etc.

A healthcare system that places greater responsibility on the individual and his/her family can encourage more responsible behavior in the usage of healthcare services and avoid wastage of scarce resources.

The family can also have more options in choosing the appropriate healthcare services in Singapore's system when compared to Hong Kong's, where the government provides heavily subsidized healthcare services and people have little choice but to accept what is given to them, regardless of the quality of service. In addition, the frequent complaint in Hong Kong is that private healthcare services often play only a supporting role to public healthcare and even when people are willing to pay more, private health practitioners may not offer more in terms of expertise and quality of service.

4.2 Sustainability

Free healthcare is not sustainable in the long run, especially in places where the taxes are low, as is the case in Hong Kong. Healthcare costs are also rising steadily over the years and the government has to face increasing expenditure. An aging population further exacerbates the problem. In recent years, Hong Kong's healthcare system is facing many problems in terms of finance and management and the chairman of the Health Authority announced in May 2005 that the healthcare system is facing a deficit of 7.3 billion Hong Kong dollars. Further, Hong Kong needs measures to better demarcate the responsibility of financing healthcare and stem the over reliance on the government.

Even though Hong Kong has implemented the MPF scheme, as part of the measure to lessen the burden of the government in providing healthcare to the elderly, overall, the MPF is not as successful a protection against retirement woes as Singapore's CPF.

Being relatively new since its induction, the amount one has to contribute to the MPF account is small (5%) in comparison to the high cost of living and the actual cost of healthcare in Hong Kong. The income earners falling below HKD 5,000 are excluded from contributing to MPF,[7] making the low income earners highly dependent on the government for healthcare in the absence of an alternative low premium insurance scheme, such as Singapore's Medishield. Upon retirement at the age of 65, MPF members can withdraw their accrued benefits and principle in one lump sum, to be used at their own discretion. Singapore used to allow its people to withdraw their CPF savings in one lump sum but has since revised the policy to include a minimum balance of SGD 80,000 and SGD 25,500[8] in the CPF and Medisave account respectively for future retirement and healthcare expenditure. This is done so that people will act more responsibly and use their CPF savings prudently after retirement.

As a rather stark contrast to Hong Kong, Singapore's healthcare system manages to run itself smoothly due to a comprehensive framework which helps people to finance their healthcare, such as Medisave, Medishield, Eldershield and Medifund.

4.3 Government Initiatives

The Singapore government classifies healthcare patients in government or restructured hospitals into six classes based on their income and/or willingness to pay for healthcare. The six classes are A1, A2, B1 (air-con), B1(non air-con), B2 and C. Only classes A1, A2 and B1 (air-con) are available in private hospitals. Government healthcare subsidies, which are funded by general taxation, are given to patients according to the class they are classified under. While Class A patients pay the full cost of medical fees, patients in classes B and C are able to enjoy government subsidies ranging from 20% to 80% of the cost. The Medishield scheme is generally able to cover a patient's healthcare costs under Class B2 or C. To avoid a premature depletion of a patient's Medisave account, caps are placed on the amount a patient can pay using Medisave according to the various classes.

Conversely, in Hong Kong, residents are normally treated as one broad classification without any differentiation due to the highly subsidized healthcare system. Theoretically, though a patient can opt to upgrade to a First or Second class ward instead of staying in the Common ward, this is hardly done because no difference in quality of treatment is perceived apart from having less people in a room. In addition, the waiting time for all patients is long due to the large demand for public healthcare services. The wait for hospital admission for surgery can be a grueling few years. To illustrate, the Chinese *Sun Newspaper* (2005, p. 1) in Hong Kong carried the front page news of a man named Mr. Cheng who suffered from spinal cord injuries but had to wait three years for treatment at a public hospital despite the urgency of his condition. Some experts warned that the patient could become paralyzed even before he was treated!

On the other hand, the Singapore government constantly upgrades public health service so that it will not fall too far behind private healthcare. For example, it was recently announced that specialists from the famous Johns Hopkins Singapore International Medical Centre will offer their services to all cancer patients at a public hospital – Tan Tock Seng Hospital. "The program is a first in Singapore in which a foreign institute is allowed to treat hospital patients receiving government subsidy" (Khalik, 2005, p. 4). Quality service at public hospitals is possible only when the patients help to bear the responsibility of healthcare costs.

The Hong Kong government's overarching concern is in reducing public expenditure on healthcare due to undifferentiated highly subsidized health services to all. Recent reforms have begun to shift some of the responsibilities of healthcare costs back to the patients. However, without an effective means of deciding who should receive more government subsidies, the system can hardly be efficient and it will generate displeasure from the lower income earners when an increase in the amount individuals have to pay for healthcare out of their own pockets is proposed across the board.

4.4 Private Sector Provision of Healthcare

In Hong Kong, private hospitals are not common and many residents choose to go to government hospitals despite the long waiting time due to nondiscriminatory, heavily subsidized healthcare for all. As stated in a 2005 report by Hong Kong Health and Medical Development Advisory Committee, private healthcare providers account for only 5% of all in-patient bed days in Hong Kong. The twelve private hospitals "usually have only a skeleton of medical staff, in particular, doctors of their own and rely heavily upon outside doctors for the admission of patients." It is not surprising then that public hospitals are the preferred choice for the people of Hong Kong.

An interestingly phenomenon in Singapore is that many patients will opt to go to private hospitals even if that means they have to incur higher costs. This is due to a few reasons. Although both Hong Kong and Singapore employers have mandatory insurance schemes for their employees, the coverage is often not enough to pay for the expensive fees in private hospitals. However, the Singapore patients frequently use their Medisave savings and/or private insurance coverage to pay for the balance. According to the International Risk Management Institute, "As of the end of 2000, there were 4 million individual policies in force that provided SGD 202 billion of insurance coverage. It is estimated that 78 percent of the Singapore population [out of a total of about 4 million] is covered by life insurance" (Kristensen & Ang, 2002). In comparison, "The number of individual life policies attained 4.6 million, covering about 69.1 percent of the population in Hong Kong [out of a total of about 6.5 million]. However there are no available statistics in the local market to reflect the actual number of policyholders" (Kristensen & Chan, 2002). Though the individual life insurance coverage for Hong Kong is quite high compared to the rest of the world, it still lags behind Singapore.

Further, Singaporeans are covered by Medishield, while their Hong Kong counterparts' MPF does not offer a similar insurance scheme. In addition, the Medishield Plus scheme is targeted at people who prefer to enjoy the facilities of Class A or B1 and higher premiums have to be paid. In late 2005, the Singapore government will further reform the Medishield scheme to allow the entry of private insurers to enhance the efficiency of the insurance plan.

Also, by placing patients into different classes, a Singaporean classified as Class A in public hospitals may like to pay a little more to be able to enjoy private healthcare, thus lessening the burden on public hospitals. There are a total of eleven private hospitals in Singapore compared to twelve in Hong Kong, where the population stands at 2.5 million more.

V. SINGAPORE SYSTEM: CONFUCIAN HEALTHCARE?
A MORAL POINT OF VIEW

Most discussions on healthcare have their focus on the economic aspects of it. However, very little notice is given to the moral dimension of the provision of healthcare. In this section, the Singaporean healthcare system will be examined vis-à-vis Confucian morality.

Confucianism does not advocate equal treatment for all and love with distinctions is always emphasized. According to Van Norden, "The doctrine of graded love states that one should have greater concern for, and has greater ethical obligations toward, those who are bound to one by special relationships, such as those between ruler and minister, father and son, husband and wife, elder and younger, and between friends" (Van Norden, 2003, p. 41). Similarly, in the area of healthcare, Confucianism does not really support a system of welfare that provides free or heavily subsidized healthcare treatment for every one in the society as this does not differentiate the people who are connected by ties. Besides, Confucianism is a family-based ethics and the primary source of healthcare should be the family instead of the state. Some interpretations of New Confucianism may take the whole state to be part of a great family and thus argue in a similar vein that the government should be responsible for its citizens' health. However, Ruiping Fan has made a cogent distinction between "big" and "small" families. "Confucians must distinguish a small family from a big family, close family members from remote relatives – all people in the world are remote relatives, and only some people are intimate relatives. It is the requirement of the Confucian principle of *ren* that one's intimate relatives receive one's preferential treatment" (Fan, 2002, p. 222).

The Confucian emphasis is that healthcare should not take a rights-based approach, but should rightfully be one based on virtues. A healthcare system based on virtues can encourage the natural flourishing of one's virtuous roots or beginnings,[9] such as a natural affection for one's children or parents. When one's closest family members fall sick, one would immediately want to employ one's best resources to their treatment and would be reluctant to leave the provision of their care to any third party. Conversely, a system that starts with one's rights to equal health treatment places the responsibility on a third party and only the government can guarantee that every one can have equal access to healthcare through public provision of it at highly subsidized rates. In addition, a rights-based approach to healthcare would inevitably be supported by the legal system, with a set of laws to prevent its violation. However, Confucians deem that laws are lowly measures to govern people's behavior and people act out of the fear of punishment but they will not be morally transformed or rehabilitated. Confucius said, "Lead the people with governmental measures and regulate them by law and punishment, and they will avoid wrongdoing but will have no sense of honor and shame. Lead them with virtue and regulate them by the rules of propriety (*li*), and they will have a sense of shame and, moreover, set themselves right" (*Analects*, 2:3; Chan p. 22). Note however, that the Confucian virtue-based healthcare system is not against welfare in the society, especially to the less fortunate. Indeed, many Confucians, such as Mencius, recommend that we extend the natural love for our family members to other non-related people in the society. "Do reverence to the elders in your own family and extend it to those in other families; show loving care to the young in your own family and extend it to those in other families..."

(*Mencius*, 1:7, p. 19). In the following, I shall provide a more comprehensive explication of the central Confucian virtue of benevolence (*ren*) and the role it plays in healthcare. More specifically, benevolence is a major concept in the Singapore Confucian system of healthcare and this concept can be used to explain the three-tiered nature of Singapore's healthcare system, but first, an explanation of the Confucian *ren*.

Ren stems from one's interaction with one's family, before extending outwards to non-family members in a network of ever-growing moral presence. We would not expect a person to be *ren* if he does not even exhibit *xiao* (filial piety) to his parents; hence, Mencius regards *xiao* as the first step towards *ren*. The early Confucians' goal was to bring harmony to the world via order in the family. *Xiao* for the early Confucians is a kind of natural affection towards one's elders in the family (biological or otherwise) and any outrage would cause extreme discomfort, such as one would feel upon the sight of foxes, flies and gnats eating one's unburied parents. "The sweat was not exuded for others to see, but was an expression of his inmost heart" (*Mencius*, 5:5, p. 125). Singapore's healthcare system is in sync with this Confucian value of *ren* and I shall explicate this through an exploration of its publicly-financed and privately-financed healthcare system, which places the main responsibility on the family, supported by the community and government.

In Singapore, healthcare is mainly funded through private savings via the family. The Medisave plan supports the expression of *xiao* because it is mainly transferable from grown-ups to their parents, between spouses and among immediate family members. In exceptional cases, Medisave money can be transferred to other close family relations. This is in line with Confucian familial ethics whereby one's closest family members have the greatest moral obligations to one another. The Medisave plan also shows the inter-connectedness of persons and the working of the special relationships because one may be simultaneously holding the roles of daughter, mother and wife and she can transfer her Medisave savings to her parents, children and husband. Meanwhile, the Medishield scheme offers protection to one's family members against expensive medical incurrences and loss of financial support in the event of serious illness or even death. This illustrates the reciprocity of care and concern among relations. This basic love for one's close family members is the first requirement of humanity. And only after one has done it well, then would one be able to extend this humane love to others.

If the family's Medisave savings cannot cover the total healthcare costs of the patient, then as a condition of *ren*, the employer will provide assistance. This reflects the Confucian ruler-minister relationship in the modern world. Employers also take care of the health of their employees by insuring them against illnesses. The family and employer constitute the two components of private financing of healthcare in Singapore.

Public healthcare provision in Singapore is mainly funded through Medifund. Medifund is a means-tested publicly funded healthcare scheme to be administered when the patient's family (and employer if applicable) cannot cover the full healthcare costs. As a last resort and a condition of a benevolent (*ren*) government, the government steps in to pay for a needy patient's healthcare. This also acts as a basic safety net in the society so that no one would be denied appropriate healthcare services.

Confucianism is a family-based ethics and the inter-connectedness of persons extends outwards in a sphere. As a result, the moral responsibility to take care of a person increases as one goes nearer the core of the sphere. This is the very principle that we can interpret the Singapore government to be working with in designing and implementing healthcare in the country.

VI. CONCLUSION

In this paper, I have spoken favorably of the Singaporean healthcare system both in economic and moral terms. A comparison is made with Hong Kong, a city rather similar historically and geographically to Singapore, but operates with a very different healthcare system. It has been shown that many areas of healthcare called up for reform in Hong Kong are precisely the ones that Singapore has done well in. Most obviously, Hong Kong's MPF scheme – modeled rather similarly to Singapore's CPF – seeks to encourage people to be more self-sufficient, especially in old age, albeit lacking in sophistication when compared to Singapore's CPF. The Hong Kong government must take greater initiatives to increase the role of the family in healthcare by helping families build a wider savings base and facilitate its use. Further, improvements in both public and private healthcare services are needed in Hong Kong to offer more choices for the people and reduce the waiting time.

APPENDIX

Table 1. Typology of Different Motivations for Sex Selection

	Singapore	Hong Kong
Healthcare expenditure (as a % of total GDP)	4.3% (2002)	5.7% (2001/02)
Public health expenditure (as a % of total health expenditure)	30.9% (2002)	57.2% (2001/02)
Private health expenditure (as a % of total health expenditure)	69.1% (2002)	42.8% (2001/02)
Private insurance (as a % of total population)	78% (est.)	69.1% (est.)
Provident fund (as a % of one's salary, total of employee and employer contributions)	25 to 33%	10%
Healthcare subsidy (as a % of total hospitalization bill)	0 to 80%	97%

Sources: World Health Organization, various Singapore and Hong Kong government, and official sources.

NOTES

[1] Free healthcare services are normally given to Comprehensive Social Security Assistance (CSSA) recipients or people exempted from payment by medical social workers.

[2] The exchange rate is about USD 1 to HKD 7.8.

[3] Various programs to combat potential health problems are initiated in schools. A successful example is the Trim and Fit program targeted at reducing obesity in children. The director of policy of the International Obesity Task Force, Mr. Neville Rigby, said recently in an interview that Singapore's programs initiated by the government to fight obesity in children seem to be the only ones that are working in the world (Low, 2005, p. 12).

[4] For more on the different class classifications, please refer to Section 4.3 on Government Initiatives.

[5] The CPF contribution rates are revisable according to the economy, age group and whether one is in the public or private sector. At one time when the economy was doing very well, the contribution rate for both employee and employer was as high as 20%. For a more detailed breakdown of CPF contribution rates, see Central Provident Fund website at http://www.cpf.gov.sg/cpf_info/On-line/contrira.asp.

[6] For more on the different class classifications, please refer to Section 4.3 on Government Initiatives.

[7] Note that although the employees earning less than HKD 5,000 per month do not have to contribute to MPF, their employers still have to contribute 5% of the incomes to the employees' MPF accounts.

[8] 1 US dollar is equivalent to about 1.7 Singapore dollars.

[9] Mencius construes a person as having Four Beginnings – humanity (*ren*), righteousness (*yi*), propriety (*li*) and wisdom (*zhi*). See *Mencius* (1999, 3:6, p. 73). This book uses the new edition classification which divides the book of *Mencius* into fourteen chapters.

REFERENCES

Chan, W. (1963). *A Source Book in Chinese Philosophy*, Princeton University Press, Princeton.

Central Provident Fund. Singapore [On-line]. Available: http://www.cpf.gov.sg/cpf_info/Publication/medisave.asp. Accessed July 2, 2005.

Department of Health. Hong Kong [On-line]. Available: http://www.info.gov.hk/dh/. Accessed June 28, 2005.

Fan, R. (2002). 'Healthcare Allocation and the Confucian Tradition,' in J. Xinyan (Ed.), *The Examined Life – Chinese Perspectives (Essays on Chinese Ethical Traditions)* (pp. 211–234), Global Academic Publishing, New York.

Gauld, R. (1997). 'Health,' in P. Wilding, A. Huque, and J. Tao (Eds.), *Social Policy in Hong Kong* (pp. 23–38), Edward Elgar Publishing Ltd., Cheltenham.

Hay, J.W. (1992). 'Healthcare: A Market Approach,' *Hong Kong Centre for Economic Research Letters*, Vol. 12 [On-line]. Available: http://www.hku.hk/hkcer/articles/v12/rhay.htm. Accessed June 22, 2005.

Health and Medical Development Advisory Committee (March 15, 2005). 'Landscape on Healthcare Services in Hong Kong' [On-line]. Available: http://www.hwfb.gov.hk/hmdac/english/dis_papers/dis_papers_lhcshk.html. Accessed June 20, 2005

Khalik, S. (2005). 'Top Docs for All TTSH Cancer Patients,' *The Straits Times*, June 1, Singapore Press Holdings, Singapore.

Kristensen, J.F. & Chan, P.W. (2002). 'The Hong Kong Insurance Market,' XL Winterthur International Co. Ltd., Hong Kong [On-line]. Available: http://www.irmi.com/Expert/Articles/2002/Kristensen03.aspx). Accessed June 2, 2005.

Kristensen, J. & Lee, A. (2002). 'The Singapore Insurance Market,' Winterthur Insurance (Far East) Pte Ltd., Singapore [On-line]. Available: http://www.irmi.com/Expert/Articles/2002/Kristensen09.aspx. Accessed June 1, 2005.

Lee, K., et al. (1998). *The Singapore Story: Memoirs of Lee Kuan Yew*, Times, Singapore.

Lee, H. (2005). 'Strong Family Ties Matter,' *The Straits Times*, February 8, Singapore Press Holdings, Singapore.

Leong, C. (2002). 'The Evolution of the Healthcare System in Hong Kong,' *Annals of the College of Surgeons of Hong Kong*, 6(3), 61–64.

Low, C. (2005). 'Obesity: Problem Among Children Growing Bigger and Health Experts Warn of a Medical Crisis,' *The Straits Times*, June 20, Singapore Press Holdings, Singapore.

Mandatory Provident Fund Authority. Hong Kong [On-line]. Available: http://www.mpfahk.org. Accessed June 26, 2005.

Mencius (1999). *Library of Chinese Classics* (Chinese-English), Hunan People's Publishing House, Hunan.

Ministry of Health. Singapore [On-line]. Available: http://www.moh.gov.sg. Accessed July 6, 2005.

Ramsay, C. (2001). *Beyond the Public-Private Debate: Access, Quality and Cost in the Health-Care Systems of Eight Countries*, Marigold Foundation Ltd, Calgary.

Tan, T.M. & Chew S.B. (Eds.) (1997). *Affordable Healthcare: Issues and Prospects*, Prentice Hall, Singapore.

The Sun Newspaper, (June 23, 2005). Hong Kong.

Tucci, J. (2004). 'The Singapore Health System – Achieving Positive Health Outcomes with Low Expenditure,' *Healthcare Market Review* [On-line]. Available: http://www.watsonwyatt.com/europe/pubs/healthcare/render2.asp?ID=13850. Accessed June 23, 2005.

Van Norden, B.W. (2003). 'A Response to the Mohist Arguments in "Impartial Caring",' in K.C. Chong, S.H. Tan & C.L. Ten (Eds.), *The Moral Circle and the Self* (pp. 41–58), Open Court, Illinois.

Wong, R. (1999). 'Hong Kong Healthcare and Finance Reform,' *Hong Kong Centre for Economic Research Letters*, Vol. 56, May-July [On-line]. Available: http://www.hku.hk/hkcer/articles/v56/healthcare.htm. Accessed June 22, 2005.

World Bank (1994). *Averting the Old-Age Crisis: Policies to Protect the Old and Promote Growth*, Oxford University Press, Oxford.

World Health Organization [On-line]. Available: http://www.who.int/en/. Accessed June 23, 2005.

CHAPTER 16

RESPECT FOR THE ELDERLY
AND FAMILY RESPONSIBILITY

*Confucian Response to the Old Age Allowance Policy
in Hong Kong*

ERIKA H. Y. YU

City University of Hong Kong, Hong Kong, PRC

I. INTRODUCTION

In face of the demographic reality of aging populations, the question "who should take care of the elderly, and how?" has become one of the most challenging policy issues for governments worldwide, and Hong Kong is certainly no exception.[1] In most cases, there are two substantial policy issues that we have to consider. First, due to the fact that health condition declines with age, an aging policy would usually involve arrangements that aim to cope with the large demand of healthcare services from the elderly. Second, generally policy issues would also cover policies to manage living expenses of the retired. Considering that medical expenses can often be an indispensable part of the living expenses of the elderly, these two policy issues are closely related. For this reason, although central concern of this paper is formulating a financial welfare policy for the elderly in Hong Kong, from time to time it also touches on the impact of such a policy on the elderly's access to healthcare services.

In view of the ever-growing aging population, the financial sustainability of providing financial assistance to elderly people through the Comprehensive Social Security Assistance (CSSA) and Old Age Allowance (OAA) is worth considering. During the decade of 93/94 to 02/03, the number of old age CSSA recipients rose significantly from 61,026 to 143,585. While the total cases of the OAA in contrast only steadily grew from 406,126 to 454,933 during the same period, these figures were rather considerable especially when compared with the number of elderly (*Hong Kong annual digest of statistics*). Besides, it should also be noted that the number of cases in the non means-tested Higher Old Age Allowance (HOAA) has

S.C. Lee (ed.), The Family, Medical Decision-Making, and Biotechnology, 197–206.

at least been double those found in the Normal Old Age Allowance (NOAA) which require means-declaration.[2] Given the low tax rate in Hong Kong and the fact that both CSSA and OAA are funded solely by the general revenue, it is hard to image how these two welfare schemes can continue to be sustainable in meeting the ever growing demand.

The reality thus imposes pressure to not only more effectively allocate resources, but more importantly, identify moral values that should guide welfare policies for elderly people. This is the case because the latter is more fundamental than the former. Indeed, unless the societal value concerning the moral obligation to take care of the old elderly is clarified and utilized to formulate the relevant policy, it would never be possible to evaluate effectiveness since there will be no clear answer as to what should be achieved with the resources. In view of this, this essay aims to shed light on the Confucian value of filial piety and put forward how the financial welfare policy for the elderly people in Hong Kong should be reshaped in light of this deeply rooted value. It reveals that although the OAA has often been regarded as a policy that promotes respect for elderly people, it is, in fact, quite a misleading perception. Instead, this paper argues from the Confucian perspective on filial piety that the OAA should be abolished in order to promote families to take care of old family members, a duty from which respect for elderly people stems.

To illustrate this case, this essay first presents an overview of the OAA policy, to give a preliminary evaluation on the objectives of the policy so as to demonstrate that there is a lack of moral vision in it. A discussion of the Confucian notion of filial duty is then followed in order to show that it is family, not society, who should take up the primary role as care giver for the elderly people. Finally, it argues that OAA should be abolished because it not only takes away resources from the elderly people who are truly in need, but also demotes the long-standing virtue of filial duty. This paper concludes with explanations on the significance of reconstructing and incorporating the Confucian value of filial piety in the aging policy of a contemporary Chinese society like Hong Kong.

II. THE OLD AGE ALLOWANCE IN HONG KONG

Whereas the OAA policy in Hong Kong, especially the non means-tested HOAA, to a large extent resembles a universal public pension system, the modest amount of the allowances suggests otherwise. In fact, rather than its official name, the OAA is more popularly known as "fruit money," which implies that it is some extra money for elderly people to spend on goods in addition to their daily necessities. "Fruit money" thus has long been recognized by the public as a sign of respect for the elderly, a moral tradition that has been highly prominent in Chinese society. However, if this is the case, then it provides little explanation for the different eligibility criteria as well as for the different amounts of allowances between NOAA and HOAA. Moreover, it is certainly even more difficult to explain why elderly recipients of disability allowances are disqualified for the OAA.

In fact, the official objectives of the OAA provide no evidence to suggest that respect for the elderly is the underlying moral value of the policy, if a clear value can be identified from them at all. According to the Health and Welfare Bureau, there are three original objectives of the OAA: (1) to provide partial financial assistance for the elderly people, (2) to encourage families to take care of their older members, and (3) to reduce the image that being old is a burden by enabling the elderly people to contribute to their family income (LegCo Penal on Welfare Service, 2001). Thus, as far as moral value is concerned, the first objective is immaterial in this evaluation. It merely outlines that the OAA is a financial measure to provide assistance for elderly people, but not a moral justification for the choices of this particular measure or age group. While the second objective has an underlying value, which holds that the family should take care of their elderly people, ironically, the OAA itself may allow families to shift at least part of this duty to the public resources. The second objective is therefore self-contradictory. This objective is also obviously at odds with the third objective, and accordingly the respective values that lie beneath them are also in contradiction, with the former promoting family to take care of their elderly people on one hand, the latter implying that it is shameful for the elderly people to be taken care by their families on the other. Clearly, the third objective also stands in the opposition with the commonly recognized traditional value of respecting the elderly.

This brief evaluation, hence, reveals that the OAA policy does not have a definite moral content in it. While the policy, as well as its divergent objectives, might be no more than some pragmatic responses to different situations identified in society, this unfortunately, overlooks the fact that public policy can often have significant influences on promoting or demoting a particular vision of morality. Public policy is never morally neutral, but should be guided by a clear moral vision in order to promote a particular moral value in society. This paper argues that in the case of Hong Kong, even though the government has often put a strong emphasis on the role of families in taking care of older family members and its aging policy appears to promote the value of respecting the elderly,[3] the OAA policy, in fact, can bring quite a contrary effect. A likely cause for this is that while the government may recognize these practices as being an important part of the traditional Chinese culture, filial piety, the underlying value of these practices, was never deliberately studied and employed to direct the policy. In view of this, some significant aspects of the virtue of filial piety will first be discussed in the next section before explaining its implications on the OAA policy.

III. THE CONFUCIAN NOTION OF *XIAO* (FILIAL PIETY)

In Chinese society, loving, respecting and supporting elderly parents is a tradition that has a long history, which can be traced back to the moral teaching of *xiao*, or filial piety, espoused by Confucianism. In the Confucian tradition, *xiao* is highly important in that it is recognized as the root of all virtues, *de* (*The Classics of Filial Piety*, Ch. I, p. 466; *Analects*, 1:2). Being *xiao*, in other words, is an essential attribute that marks human beings from other animals as possession and practice of virtues are central to humanity. From the Confucian perspective, individuals are all by nature socially related, and virtues are significant because they enable individuals to relate to others in a harmonious fashion. Among various virtues, *ren*, which can be regarded as loving other humans, is the cardinal one (*Analects* 12:22).[4] *Xiao*, which prescribes the moral requirements of children in relation to their parents, is recognized as the most fundamental virtue not only because everyone must be born into a parent-child relationship, but more importantly individuals usually experience and learn *ren* first from their parents due to the natural love of parents to their children, and consequently, it is also most natural and sensible for individuals to reciprocate *ren* first to their parents. *Xiao* is thus a practice of the virtue *ren* from children to their parents, and this is so significant that in *Mencius* (4A:27) it is pointed out that, "the content of *ren* is the serving of one's parents" (see also 7A:15). In short, *xiao* is of particular importance for Confucians because one can never be virtuous without being *xiao*.

As a practice of *ren*, *xiao* can certainly entail a wide range of conducts; while for the common people, it primarily concerns the duty to serve one's parents with love and reverence.[5] According to the Confucian classics, although taking care of one's parents is unquestionably the service which filial adult children must give to their parents, merely providing care to one's parents is still far from being *xiao* in the fullest sense.[6] Indeed, the duty to take care of one's parents in terms of material goods is seldom emphasized when *xiao* is being defined (if it should deserve to be recognized as *xiao* at all) for it is such a bare minimum that Confucians would hardly consider the need to point it out.[7] Instead, whenever looking after the livelihood of elderly parents is seen as an important matter, the subject concerned is usually the state or ruler, who must examine his role in ensuring his people's ability to discharge this minimum duty.[8] *Xiao*, therefore, when prescribed as an individual's duty as a child, must always require more than simply taking care of the livelihood of one's elderly parents. Yet, for a state, this minimum sense of *xiao* is of importance as it serves to remind us that a state should ensure and promote families in fulfilling the duty to look after the livelihood of their elderly members.

While Confucians assert that respectful care for elderly people should be first and foremost the duty of adult children, this by no means implies that the virtue should not extend to those outside of one's family. Instead, as a practice of *ren*, one should eventually extend love and respect to all the elderly. However, from the Confucian perspective, it is crucial that this is practiced with gradation, that is, one must first love and take respectful care of one's own elderly parents to the greatest extent possible before one extends such practices to other elderly. This is because one's roles as parent and child is always more primary than one's role as a members of society.[9] Thus, according to Mencius, even though a true King should practice *ren* to all his people, he must *first* "treat the elderly of your own family in a manner befitting their venerable age," *then* "extend this treatment to the elderly of other families" (*Mencius* 1A:7). In contrast, it can also be observed from the *Classic of Filial Piety* that if anyone does not love one's parents but others, this would be seen as "rebelling against virtue" (*bei de*), or if anyone does not revere one's parents but other men, this would be seen as "rebelling against propriety" (*bei li*) (*The Classics of Filial Piety*, Ch. IX, p. 479).[10] Therefore, based on the virtue of *xiao*, elderly people should be taken care of by their adult children, for this is the moral duty of the latter. Public resources, on the other hand, should be directed to those elderly people who are in need but who do not have younger family members to support them. In this way, people cultivate their virtues first by being *xiao* to their own elderly parents, then by extending *ren* to other elderly people in the community.[11]

IV. THE IMPLICATIONS OF *XIAO* ON THE OLD AGE ALLOWANCE POLICY

While the OAA policy may often be credited for its role in reflecting and promoting the traditional value of respecting the elderly, the above discussion on the Confucian value of *xiao* reveals an essential moral gauge that this common perception can be measured against. This section argues that based on the time-honored virtue of *xiao*, the OAA is indeed a policy bare of moral worth and therefore should be abolished.

First of all, the OAA policy may serve to justify undutiful adult children by endorsing the notion that the livelihood of the elderly is merely a universal entitlement. According to the Confucian teaching of *xiao*, looking after the welfare, or at least the livelihood, of elderly parents is a necessary family obligation of dutiful adult children. This is not an entitlement because it should be a natural reciprocation of love *from children* to their parents. In fact, it should not even be compared to an entitlement because being served by children with love and reverence and being able to secure the well-being of one's parents, are essential attributes of the flourishing life. This also explains why in the Chinese culture it is always an honor for the elderly to be dependent on their children but a shame for adult children who fail to support their parents. The OAA, on the other hand, being non means-tested and which applies universally to all elderly people in Hong Kong, may "at best" misdirect the public to perceive that society is the primary institution in taking care of the elderly people whereas family is only the secondary one. It may mislead or even justify adult children to believe that they only have a partial duty to take care of the livelihood of their parents, since society can always take up a part of it, regardless of the competency of the adult children in fulfilling their filial duty.[12] As the same judgment applies to healthcare services, it leads children to think that they can always rely on the government to pay most of the bills for the medical expenses of the elderly. Thus, instead of trying to afford the best healthcare for their elderly parents, children from different financial backgrounds all have a common justification for putting their elderly parents in the public healthcare system, despite the tremendously long waiting time, highly limited consultation time and lower quality of drugs. Consequently, the pressure put on the public health system grows along with the aging society, resulting in a lower quality of service. At worst, the elderly may even be left with nothing from their adult children except for the modest amount of OAA given to cope with all their daily expenses. Considering the fact that the lives of these elderly people must be unquestionably difficult, it is shameful to suggest that the OAA, even in this case, could be a sign of respecting the elderly. Moreover, in the face of such financial difficulty, it is likely that the elderly would avoid seeking medical treatments even if they are in great need of them.[13]

Certainly, this is not to suggest that society should not take care of the elderly whatsoever. As previously mentioned, Confucians would be delighted to see individuals being able to extend and practice their *ren* to the elderly outside of their own family. Moreover, it can also be valid to argue that society should have a moral duty to look after the elderly due to their vast contributions in the past, a moral ground that indeed has often been employed to justify the OAA policy. Nonetheless, it must be noted that there should always be a clear order between societal and family obligations, with the latter more fundamental than the former as explained. In other words, even though the elder people may have given considerable contributions to society in their early years, we must not forget that they should have given away even more to their own family. Consequently, the family should still be their primary care taker unless elderly people do not have this resort. Additionally, in Hong Kong, the CSSA, but not the OAA, is established to safeguard the livelihood of these unfortunate elderly people. While the societal obligation discharged by the CSSA policy is obviously not limited to the elderly people, the higher amount of assistance given to the elderly may nonetheless serve as a symbol of respect for the elderly.

The OAA policy, by misarranging the proper priority of societal and family obligations, may also serve to induce and reinforce the image that being old is a burden even though its objective suggests the exact opposite. As discussed, in light of *xiao*, the family obligation of serving elderly parents should never be perceived as a burden. In fact, one would not even view taking care of elderly people as a burden if they do so as a result of extending their *ren*. Nonetheless, the perception that supporting the old is a burden results when individuals are discouraged to cultivate their virtue of *xiao* by fulfilling one's family obligation, but at the same time are forced to support other elderly in the name of societal obligation. After all, if individuals, as children, are demotivated to look after their own elderly parents who always love and care about them, then why should they be obligated to support other elderly people who are no more than strangers to them? Why should they not regard these elderly people as being a burden to them? The OAA policy can lead to, in short, the Confucian notions of *bei de* and *bei li*. Further, discriminating against the elderly would hardly be a respectful and justifiable act to them.

Surely, in addition to abolishment of the OAA policy, measures must also be taken by the government to assist those families that are not capable of providing full support for their elder members. Instead of providing direct financial assistance to the elderly like the case of OAA, the government should initiate other measures such as facilitating employment and providing more tax allowances for adult children who take care of or even live with their parents. Even if financial subsidy is needed to cover costs, it must still be means-tested by taking the elderly parents and their children, or perhaps their children's families if applicable, as a unit for assessment. This is important because it not only ensures a decent livelihood for the elderly, but it also ensures children are committed to fulfill their filial duties. In this way, the proper order of familial and societal obligations would not be upset.

Another issue that the government should consider is whether to enact a relevant legislation, as a last resort, to ensure that capable children take care of their parents.[14] After all, while adult children are not asked to support other elderly people before their own, this does not necessitate that they will take care of their elderly parents. Although it should be stressed that filial duty should not stem from legal obligation, the law may serve to convey a clear message on the significance of this traditional moral value and demonstrate the societal resolve to enforce it.

All in all, the OAA should be abolished for it is a highly ineffective policy in light of the value of *xiao*. It not only fails to promote the time-honored value, worse still it demotivates individuals to cultivate the virtue and even justifies children to not comply with their filial duties. Elderly people, consequently, may no longer be a subject for respect but are perceived as a burden. Besides, unnecessary expenditures on the non means-tested OAA also divert resources from the elderly who are truly in need.

V. CONCLUDING REMARKS

While the question "who should take care of the elderly and how?" is undoubtedly a universal policy challenge for governments, the answer to it is not. This is because apart from taking into consideration the general principles advocated by international institutions, aging policies must also be guided by a moral vision, which is often specific to the local context. The latter is of substantial importance because policies can have a considerable influence on social morality by promoting or demoting particular moral values.

This paper reveals that in Hong Kong, although the aging policies in place do take some local conventions into account, the underlying moral values and moral implications of these policies may often be overlooked, and the OAA policy is a case in point. Despite the common perception that the OAA policy plays a role in strengthening the traditional Chinese value of respecting the elderly, this paper argues that the OAA policy indeed can only bring the reverse when we consider the Confucian value of *xiao* upon which respect for the elderly people is founded. Hence, the major argument in favor of keeping the OAA, namely, that it serves to express respect for the elderly, is implausible. Furthermore, although there may be real cases where some elderly people for whatever reasons are reluctant to apply for the means-tested CSSA and consequently rely solely on the OAA for livings, we have shown that this, in fact, only provides us with a stronger reason for abolishing the OAA policy. Finally, some are concerned that the abolishment of the OAA policy is no more than an excuse for the government to cut the benefits of elderly people in order to alleviate its budget deficit. Such concerned can be alleviated if the government can guarantee that the total budget spent on elderly welfare is not going to be reduced.

In Chinese society, the family has always served as a non-substitutable primary caretaker of elderly people. This is reinforced by the deeply-rooted Confucian notion of *xiao*. While there is the undeniable reality of the ever decreasing birth rate in Hong Kong and the fact that a more individualistic approach, such as MPF, may be pragmatically needed as a part of the aging policy, this must not undermine the importance of considering Confucian virtues and their roles when constructing health policies that apply to the elderly. First and foremost, these virtues serve a key role in educating, both parents and children about the parent-child relationship necessary to cultivate virtues and to live a flourishing life. In the arena of public policy, social values and individual decision-making are often interrelated; eventually, this may improve the low birth rate in the city. It is important to note that given the prospect increased life expectancy, no healthcare or social security systems can be sustainable if the elderly dependency ratio continues to increase. Second, such system would be fair to all parents, who have made enormous sacrifices to bring up their children. They deserve more love and care from their children than other elderly persons. Finally, this may also keep the modern society from overemphasizing institutionalization and efficiency when it comes to matters concerning human relations. Taking care of the elderly involves things like love, respect and a sense of belonging, which no social security system can assure. The virtue of *xiao* serves to remind individuals about the essential interdependent nature of man as they care for their elderly parents who used to care for them.

NOTES

[1] According to the Census and Statistics Department, it projected that the percentage of the population group elderly 65 or above would rise from 11% in 2001 (the base year) to 24% in 2031, and the elderly dependency ratio from 382 in 2001 to 562 in 2031. (Demographic Statistics Section, Census and Statistics Department, 2002 p. 9).

[2] In the financial year of 91/92, the cases of HOAA and NOAA were 284,526 and 126,694, respectively, and in the year 02/03, the numbers of recipients of HOAA and NOAA were 348,012 and 106,921, respectively (Social Welfare Department, various years). The expenditure on the OAA payments was amounted to HK$1,847.8 millions and HK$3,574.0 millions in 91/92 and 02/03, respectively (Census and Statistics Department, various years).

[3] For instance, in his 1997 Policy Speech, the Chief Executive mentioned that "caring for the elderly is the responsibility of every family," *1997 Policy Speech: Building Hong Kong into a New Era* (1997, para. 114).

[4] It should be noted that apart from love, *ren* is also often understood as benevolent, kindness, etc. This is because *ren*, as the most significant quality that men should cultivate, is a highly complicated notion and its exact meaning is often context specific.

[5] From the *Classics of Filial Piety* (Ch. I, pp. 465–467), it can be observed clearly that *xiao* can imply a wide range of conducts because in order to attend the end of filial piety, one must establish one's character and make one's name famous in future ages so as to glorify one's parents. This, in other words, suggests that the requirement of being *xiao* can go as far as the possession of all supreme qualities. In Chapter II–V, it can also be observed that individuals in different positions may have different requirements for being recognized as *xiao* (pp. 467–472).

[6] Chapter X of the *Classics of Filial Piety* (p. 480) may provide the best summary of the content of serving one's parents, "the service which a filial son does to his parents is as follows: In his general conduct to them, he manifests the utmost reverence. In his nourishing of them, his endeavor is to give them the utmost pleasure. When they are ill, he feels the greatest anxiety. In mourning for them (dead), he exhibits every demonstration of grief. In sacrificing to them, he displays the utmost solemnity. When a son is complete in these five things, (he may be pronounced) able to serve his parents."

[7] See, for instance, *Analects* (2:7) & (2:8).

[8] For example, see *Mencius* (1A:7).

[9] This can be observed from the well-known though highly controversial example from *Analects* (13:18) where Confucius points out that fathers cover up crimes committed by sons or sons cover up crimes for their fathers is more justified than sons report the crimes of fathers so as to serve as a responsible citizen. While this case may have often been condemned in modern society where rule of law is considered to be highly important, it must take into consideration that many contemporary legal systems also exempt family members from testifying against each other.

[10] The possible meanings of other men are more inclusive than simply other elderly, it indeed can be any human beings. Nonetheless, the key point is that one should always love and respect one's own parents first.

[11] It can be observed that the Confucian notion of *xiao* demonstrates a sharp contrast with some Western views on filial obligation, such as Norman Daniels who argued that filial duty should not be a corollary of the self-imposed parental obligation, and children should have no obligation to reciprocate the good done to them by their parents. See Daniels (1988). For Daniels, the moral requirements between parents and children are no more than a result of consents by *independent individuals*. However, Confucians could not take this view because it overlooks the natural duties which all naturally *related persons* should discharge and the importance of this for the flourishing life. While Daniels might have recognized that individuals might want to be good parents because this could contribute to a more meaningful life for them, he has missed the Confucian insight that being a filial child is also highly significant for the good life since it allows individuals to realize their natural desire to be related through the most fundamental and natural parent-child relationship. Hence, it should be reminded that although reciprocity is one of the justifications for *xiao*, it is not the only one. This is also the reason why in some exceptional cases where even though parents fail to be kind, Confucians would still advocate the children to be filial. A classical example is the case of Shun, a sage whom Confucius often mentions as a role model of perfect character. See *Shiji*, Vol. 1. Indeed, just as the general cases show that good parents can usually cultivate filial children, the story of Shun demonstrates that filial children can also cultivate the kindness of parents. And this can also explain partly why Confucians believe that relationships guided by proper conducts are critical for virtues cultivation. Moreover, the Confucian notion of *xiao* can also illustrate, in contrast with Daniels's argument against "privatization" of social obligations concerning the elderly, why taking care of the elderly should not primarily be regarded as social obligations. This point will also be explained in more details in the next section.

[12] Although the NOAA requires income and asset declaration from the elderly applicants, it does not require the family of the elderly to do so. This means-tested measure therefore never says anything on the financial ability of the family in supporting their parents. Besides, as outlined before, the recipients of the HOAA, which is completely non means-tested, are in much higher numbers than that of the NOAA.

[13] Even though all OAA recipients must be permanent Hong Kong residents and therefore their fees for the public health services have already been highly subsidized by the government, the service charges can still occupy a significant proportion of their OAA. For instance, the accident & emergency service costs HK$100 per attendance, in-patient service requires HK$68 per day and general out-patient also charges HK$45 per attendance (drugs included). The cost for specialist outpatient service certainly can go much higher since it takes HK$10 for each drug item in addition to an attendance fee. For more details on the public health services fees, please see the Hong Kong Health Authority (2005).

[14] Similar practice can be found in the Chinese Marriage Law in China. See Wang (1999, p. 245). Recently, a lawyer in China has even proposed a law that is specific to filial duty. See "Should Filial Duty Become Law?" (2004).

REFERENCES

Anonymous (2004). 'Should Filial Duty Become Law?,' *Beijing Review*, October 28, pp. 44–45.

Census and Statistics Department (yearly). *Hong Kong Annual Digest of Statistics*, Census and Statistics Department, Hong Kong.

Confucius (1976). 'The Classics of Filial Piety (The Hsiâo King),' in J. Legge (trans.), *The Sacred Books of the East: The Texts of Confucianism*, Vol. 1, Gordon Press, New York.

Confucius (1979). *Analects*, D.C. Lau, (trans.), The Chinese University Press, Hong Kong.

Daniels, N. (1988). *Am I My Parents' Keeper*, Oxford University Press, New York.

Demographic Statistics Section, Census and Statistics Department (2002). *Hong Kong Population Projections, 2002–2031*, Census and Statistics Department, Hong Kong.

Hong Kong Health Authority (2005). 'Fees and Charges for Public Healthcare Services' [Online]. Available: http://www.ha.org.hk/hesd/nsapi/drivernsapi20.so/?MIval=ha_visitor_index&intro=ha%5fview%5ftemplate%26group%3dOSR%26Area%3dFNC. Accessed June 27, 2005.

LegCo Penal on Welfare Service (July 2001). 'Financial Support for Older Persons' [Online]. Available: http://www.hwfb.gov.hk/hw/english/archive/legco/W_9_7/fsope.HTM. Accessed May 14, 2005.

Mencius (2003). *Mencius*, D.C. Lau, (trans.), The Chinese University Press, Hong Kong.

Sima, Q. (1959). *Shiji*, Vol.1, Chung Hwa Book Co., Peking.

Social Welfare Department (May 25, 2005). 'Figures on Social Security' [Online]. Available: http://www.info.gov.hk/swd/html_eng/ser_sec/soc_secu/index.html. Accessed May 30, 2005.

Social Welfare Department (2004). *Social Security Allowance Scheme*, Hong Kong Logistic Department, Hong Kong.

Social Welfare Department (yearly). *Social Welfare Departmental Annual Report*, Census and Statistics Department, Hong Kong.

The Monetary Provident Fund Schemes Authority (June 10, 2005). *The Monetary Provident Fund Schemes Authority website* [Online]. Available: http://www.mpfahk.org/index.asp?langid=1. Accessed June 10, 2005.

Wang, Q. (1999). 'The Confucian Filial Obligation and Care for Elderly Parents,' in R. Fan (Ed.), *Confucian Bioethics*, Kluwer Academic Publishers, Boston.

World Bank (1994). *Averting the Old Age Crisis: Policies to Protect the Old and Promote Growth*, Oxford University Press,Oxford.

1997 Policy Speech: Building Hong Kong into a New Era (1997) [Online]. Available: http://www.policyaddress.gov.hk/pa97/english/patext.htm. Accessed June 2, 2005.

CHAPTER 17

IS SINGAPORE'S HEALTH CARE SYSTEM CONGRUENT WITH CONFUCIANISM?

JUSTIN HO

Rice University, Houston, Texas

All countries face the task of implementing a health care system that is financially sustainable and morally defensible. In "Confucian Health Care System in Singapore: A Family-oriented Approach to Financial Sustainability," Kris Su Hui Teo argues that Singapore's health care system has done well in balancing and executing these dual tasks, having developed a three-tier model that is both economically sound and defensible within a larger framework of Confucian values.

In particular, it has been claimed that Singapore has developed a health care system that limits: (1) the moral hazards involved in health care entitlement systems (e.g. governmental as well as private insurance schemes); (2) the political hazards involved when governments establish a health insurance scheme which allows politicians to promise further entitlements that will need to be funded (if possible) in the future and (3) the demographic hazard of making the fundability of health care dependent on sufficient young workers to pay for the health care of the unemployed and the aged. This short essay does not attempt to examine whether Singapore's health care is as economically successful as many have claimed. Rather, I explore whether Singapore's system is, in fact, as consistent with Confucianism as Teo argues. This essay examines whether Singapore's approach to health care is contrary to the Confucian view of an ideal society. In addition, I will show that even if Singapore's health care system does conflict on some points with this ideal, if such a health care system serves an overriding or higher Confucian moral concern, such as the harmony of the state and the universe, then such measures might nevertheless be justified according to Confucianism. However, prior to addressing this issue directly, it is necessary first to lay out some of the central tenets of Confucianism in order to better understand both the possible moral shortcomings as well the positive features of Singapore's health care system.

S.C. Lee (ed.), The Family, Medical Decision-Making, and Biotechnology, 207–215.
© 2007 *Springer.*

I. SOME CENTRAL TENETS OF CONFUCIANISM

Confucianism is a teleological theory of morality; it claims that the end of all human action is to bring about the harmony of the universe (Chan, 1963). This ideal is articulated in the *Doctrine of the Mean* where it is explicitly argued that "when the equilibrium and harmony are realized to the highest degree, heaven and earth will attain their proper order and all things will flourish" (Chan, 1963, p. 98). This view is also prominently displayed in the first chapter of the Confucian classic, *The Great Learning*, where the following doctrine about ideal human action and the purpose of such action is outlined:

(1) Investigating Things
    ↓leads to
(2) Extending knowledge
    ↓leads to
(3) Making the will sincere
    ↓leads to
(4) Rectifying the heart-mind
    ↓leads to
(5) Cultivating the person
    ↓leads to
(6) Regulating the family
    ↓leads to
(7) Governing the state well
    ↓leads to
(8) Bringing harmony to the universe.

According to this account, each of the subsequent steps is dependent for its realization upon the completion of the previous step. Specifically, the harmony of the universe can only be obtained by bringing about the harmony of the society in which one lives, and this precondition can itself only be realized by making a society's families harmonious, and this in turn can only occur through personal cultivation. Therefore, the harmony of the universe can only be achieved by cultivating oneself.

Now to cultivate oneself is to become a virtuous person or a sage. This is illustrated in the passages: "as a virtue, nothing is greater than equilibrium and commonality" (*Analects* 6:29) and "The serious man is one who cultivates himself with seriousness...so as to give all people security and peace" (*Analects* 14:4). In Confucianism then, a sage is the full embodiment of what it is to be a person. Crucial here is the complex notion of *ren*. The word *ren* has two different meanings in the *Analects*. *Ren* refers to "the consummation of personal excellence," as well as the virtue of benevolence (Van Norden, 2002). This is evidenced by the fact that there are passages where:

(1) *ren* is listed as one virtue among others (The Master said, 'The wise are free from perplexities; the virtuous [*ren*] from Anxiety; and the bold from fear.' *Analects* 9:28);

(2) *ren* describes something that includes other virtues such as courage ('When the love of superiority, boasting, resentments and covetous are repressed, this may be deemed perfect virtue [*ren*].' *Analects* 14:1);

(3) *ren* is described as love for others (1. Fan Ch'ih asked about benevolence [*ren*]. The Master said, 'It is to love all men.' (*Analects* 12:22).

In this paper, I will use *ren* (a) to refer to the first meaning and *ren* (b) to refer to the second meaning. A person who is *ren* (a) (Mencius 7B:16) is a moral individual who has fully cultivated all of the separate virtues to the highest degree (Schwartz; Waley; Van Norden).[1] Such a person will perform *yi* or appropriate actions and is also benevolent (*ren* (b)). As Hui (1999; 2004) has noted in his work on Confucian bioethics, *ren* is a relational term. The ideograph for *ren* is composed of two characters "man" and "two", "denoting that whatever else the term may mean, its meaning is intended to be accomplished through human relationships" (Hui, 2004, p. 35). A person who is *ren* (a) has perfected his relations with others, which is a state described by Confucius in various passages in the *Analects*. For example, he is reported to have characterized being *ren* (a) in the following way: "A man of humanity [*ren* (a)] wishing to establish his own character, also establishes the character of others, and wishing to be prominent himself, also helps others to be prominent" (*Analects* 6:28). Mencius builds on this notion, noting that an adult person tends be situated in a number of human relationships. S/he may be a husband or wife, a father or mother, a subject or ruler, and may be an elder or a youth. A person who is *ren* (a) expresses the type of affection that is appropriate given one's role in a particular relationship (Mencius 3A:3). Furthermore, in response to several passages in the *Analects* which suggest that one should always be pious to one's parents (*xiao*) (e.g. *Analects* 2:5, 2:6, 4:19), many scholars who followed Confucius have taken this to mean that one ought to extend more love to those persons who stand closest in relation, as those are the persons who are most directly affected by one's expression of love.

According to Confucianism, *li* plays a central role in one becoming a *ren* person. In the *Analects*, *li* is used to refer to (1) religious rituals (*Analects* 2:5), (2) rules of etiquette (*Analects* 9:3) and (3) proper behavior in general (*Analects* 4:16). *Li* should not only guide one's decisions in ethical matters, but rather the whole of one's life should be lived in accordance with *li*. Confucius told one of his disciples: "do not to look at what is contrary to *li*, do not listen to what is contrary to *li*, do not speak of what is contrary to *li*, and do not make a movement of what is contrary to *li*" (*Analects* 7:1), as *li* helps one learn how properly to conduct one's relations with others, which in turn cultivates virtue in oneself, as well as others. The latter claim follows because Confucianism holds that virtuous persons can have the effect of making others virtuous as well (*Analects* 2:1). Moreover, one can only become *ren* through *li* (*Analects* 12:1).[2] To expound on this last point, there is something mysterious about how the proper practice of *li* transforms one into a *ren* person. When asked about how rituals such as Ancestral Sacrifice are able to bring about positive change in oneself, Confucius said, "I do not know. Anyone who knew the explanation could deal with all things under Heaven as easily as I lay this here;

and he laid his finger upon the palm of his hand" (*Analects* 3:11). One possible hypothesis might be as follows. *Li* in a very general sense refers to proper form. Some later Confucians have held the view that a person is comprised of a body (*ti* or *shen*), heart-mind (*xin*), and vital force (*qi*) (Ni, 1999). One might argue that through the practice of *li*, all of these aspects of a person are brought in equilibrium or harmony with one another, and consequently, a person becomes harmonious or virtuous.

This general ethical-ontological framework deeply informs Confucian political theory to the effect that a society should ultimately be organized around *li*. That is, it should be structured by rightly ordered rituals. A just state is a harmonious state, a quality only achievable through the moral cultivation of its citizens through their recognizing the centrality of ritual. Contrary to Aristotle (*Nicomachean Ethics*, 1999), however, Confucianism claims that a government should not attempt to facilitate this process through laws and policies, but rather through *li* and the natural moral inclinations of its people.

(1) If the people be led by laws, and uniformity sought to be given them by punishments, they will try to avoid the punishment, but have no sense of shame.
(2) If they be led by virtue, and uniformity sought to be given them by the rules of propriety, they will have the sense of shame, and moreover will become good (*Analects* 2:3).

The "sense of shame" resulting from violations of duty can be interpreted as evidence for the proper internalization of duty, which works to prevent persons from performing immoral actions. The rationale behind this claim seems to be the following:

(1) Moral people act appropriately.
(2) Moral cultivation can only be achieved through one's relations with others and ultimately begins with one's family.
(3) Therefore, it is or should be the family's responsibility to cultivate the morality of citizens.

It follows from this rationale that the government should refrain from issuing positive laws whenever possible.

II. EVALUATING SINGAPORE'S HEALTHCARE SYSTEM

With the above framework in mind, let us now turn to the primary objectives of this paper: namely, evaluating the extent to which Singapore's healthcare system is consistent with central Confucian commitments.

To begin with, Teo is right to note that by making the individual and his or her family the primary bearers of choice and responsibility in health care financing and decision-making, Singapore's healthcare system creates space for the central Confucian values of *ren* (a) and *xiao*. Moreover such policy is in accord with the Confucian account recognition of the family as morally prior to the individual. Singapore's health care system in particular is congruent with *ren* and *xiao* in that money from medical savings accounts are transferable to immediate family members. By encouraging families to provide for their members' health care needs, persons are able to express the type of affection that is appropriate, given one's role in a particular relationship, and which is necessary to be a *ren* (a) person. More importantly, such actions promote the harmony of the family, which in turn contributes to the harmony of the state and the universe, as it is only when all of the members of the family are expressing their love appropriately to one another that the family as a whole is made peaceful. As noted in other essays in this volume (e.g. Fan, 2007; Engelhardt, 2007) the proper relation of family members within the reality of the family is crucial.

However, Singapore's health care system also provides government-subsidized health care to the medically indigent, i.e., those persons who do not have the financial resources to pay for healthcare. This is inconsistent with Confucian ideals. As Fan (2004) interprets these ideals as laid out by Mencius, the best community is a well-field district in which people learn the virtue of *ren* by taking care of each other:

> In the field of a district, those who belong to the same nine squares render all friendly offices to one another in their going out and coming in, aid one another in keeping watch and ward, and *sustain one another in sickness*. Thus the people are brought together to live in affection and harmony (Mencius, 3A:3:18) (my italics).

In this passage, the local community and not the government is held to bear primary responsibility for providing care (and implicitly health care) to persons. Not only does Singapore's state intervention collide with the Confucian focus on limiting governmental intervention, it may in some cases undermine or discount the centrality of the family. It might be argued that, according to Confucianism, the government should refrain from providing such aid.

Secondly, one might critically note that the mandatory character of medical savings accounts might be contrary to the Confucian ethical and political ideal of a limited government. This concern is strengthened when one considers the fact that the government has implemented a number of other laws intended to curtail spending on government-subsidized healthcare services. In his well-documented article, "Medical Savings Accounts in Singapore: A Critical Inquiry," Barr (2001) notes that in addition to paying for all of the healthcare expenditures of the medically indigent, the government assigns hospital patients to classes of service based on their income and/ or willingness to pay for healthcare. Depending on the class to which they are assigned, patients in government hospitals may be charged nineteen to eighty percent of the total costs, with the government subsidizing the rest while private hospitals charge patients 100 percent of the costs (Barr, 2001). Barr also points out that a review of documents composed by the Ministry of Health shows that in order to control government spending on healthcare subsidies, Singapore has (1) regulated the introduction of technology and access to medical specialists in government hospitals; (2) introduced a predetermined rate of subsidy for these institutions; (3) ensured that the number of specialists number no more than 40 percent of the medical profession (1, 2, and 3 are reported in a document by the Ministry of Health entitled "Report of the Cost Review Committee-Response of the Ministry of Health"); (4) introduced price caps on all medical services delivered in government hospitals (Low, 1998) and (5) restricted the number of government hospital beds (Massaro and Wong, 1995). In addition, there are limits placed on the sorts of services for which medical savings accounts and catastrophic insurance can be used.

The problem can be summed up as follows: rather than attempting to cultivate the moral character of its citizens through *li* and adhering to the Confucian model of an ideal society, Singapore has instead imposed by law a number of constraints (1) forcing employees and employers to contribute to medical savings accounts and (2) limiting their spending on government services so as to ensure the financial wellbeing of the country.

III. SOME POSSIBLE REJOINDERS

The force of the second criticism (i.e. the mandatory character of Singapore's healthcare policy) can be weakened. As was already noted, according to Confucianism, the end of all human action is to bring about the harmony of the universe as a whole and the harmony of the state is a necessary condition for the realization of this end. In this light, actions that promote the harmony of any given society are generally moral actions that ought to be pursued and encouraged. Though Confucianism holds that the government should refrain from passing positive laws, in that such measures do not promote moral cultivation and therefore, fail to bring about the harmony of the state, contemporary health care may be an exception. If human actions should always be directed toward the higher goal of bringing about the harmony of the universe, it can be argued that Singapore's health care system is in line with Confucian ethics because it promotes the harmony of the state, which in turn contributes to the harmony of the universe. By requiring people to contribute to and maintain a medical savings account, and by curtailing government spending, Singapore is able to maintain a healthcare system that is financially stable and which encourages family solidarity. Singapore's approach may be that which is necessary to ensure the long term survival of the country as a whole. If such polices were not implemented, the country might face economic difficulties so that the policy as a whole might be marked by dissension if not chaos.

In further support of this claim, consider the story of Guan Zhong. Guan Zhong violated the Confucian value that one should sacrifice one's life if necessary for the sake of upholding *ren* and *yi* when after his master Prince Jiu was assassinated by Prince Xiaobai, he chose to live through serving the assassin (Lo, 1999). His actions were justified according to Confucius because:

> Guan Zhong helped Duke Huan to become leader of the feudal lords and to save the Empire from collapse. To this day, the common people still enjoy the benefits of his acts. Had it not been for Guan Zhong, we might well be wearing our hair down and folding our robes to the left. Surely his was not the petty faithfulness of the common man or woman who commits suicide in a ditch without anyone taking notice (*Analects* 14:16–17).

From Confucius' remarks here, it is plausible to infer that it is permissible to violate "lesser" Confucian values in order to promote "higher" Confucian values such as the harmony of the state and the universe.

For many Confucians, such a response may be unsatisfactory when considering laws intended to curb the use of governmental health care services. In this vein, one might argue that such laws are the result of the government subsidizing health services provided by government hospitals and supplies free health care to the medically indigent, which as we have noted, is contrary to the Confucian account of an ideal society. Had the government not acted contrary to the Confucian ethos in the first place, then there would be no need to put into place such controls on expenditures. Furthermore, many Confucians might also argue that eliminating such subsidies and aid is also in accordance with the aim of promoting the harmony of society and the universe. They could point out that such measures would also result in a healthcare system that is more financially stable, and which, in turn, would better ensure the long term survival of the country as a whole, in that the government would no longer have to worry about funding and controlling rising health care costs that may result from an aging population. Such persons might keenly note that currently the public sector provides 80% of inpatient care (Teo, 2007), and that Singapore has yet to face the costs of an aging population (Barr, 2001). Critics of governmental support of health care for the indigent (in contrast to local community support of those in need) would likely also point out that eliminating subsidies would dissolve the first of the criticisms that we considered, namely that giving free medical care to the poor is contrary to the Confucian notion of an ideal society, which relies on local government support.

Nevertheless, in rejoinder, one can argue that by offering subsidies and aid to the medically indigent, the government prevents dissension and possibly even revolt from occurring that would over time weaken the foundations of society. If that were indeed the case, then such measures would be fully warranted. Ultimately, further analysis informed by empirical data is needed before one can determine more conclusively which of these positions ought to be accepted.

IV. CONCLUDING REMARKS

In summary, there are considerations that allow one to argue that Singapore's health care system might be either contrary or in accordance with the tenets of Confucianism. For instance, one can argue that Singapore's health care system is, in fact, consistent with a Confucian ethos if it is the case that such a system best promotes such core Confucian values as the harmony of family, society and the universe. Alternatively, one can argue that Singapore's healthcare system is inconsistent with such an ethos if (1) it can be shown this is not in accordance with the Confucian model of an ideal society and (2) that an alternative system better takes into account Confucian moral and ontological commitments (e.g. with regard to the harmony of the family) and would, therefore, better promote the harmony of society and the universe.

This essay sought only to evaluate Singapore's healthcare system from a Confucian perspective, rather than actually to endorse or find fault with the overall merits of Singapore's actual health care system. Nevertheless, for those persons for whom Confucian values play a central role in their value structures, the questions and issues raised should be a deciding factor in their decision whether to embrace or reject Singapore's healthcare system.

ACKNOWLEDGEMENTS

The author would like to thank H. Tristram Engelhardt, Jr., and Jeremy Garrett for many helpful comments.

NOTES

[1] See Ivanhoe's *Confucian Moral Self Cultivation* for a more in-depth analysis of how one becomes a Sage (Ivanhoe, 2000).
[2] See Kwong-Loi Shun (2002) "Ren and Li in the *Analects*" for an excellent discussion of the relation of *ren* and *li*.

REFERENCES

Aristotle (2002). *Nicomachean Ethics*, S. Broadie and C. Rowe (Eds.), Oxford University Press, Oxford.
Barr, M.D. (2001). 'Medical Savings Accounts in Singapore: A Critical Inquiry,' *Journal of Health Politics, Policy and Law*, 26 (4), 709–726.
Chan W.T. (1963). *A Source Book in Chinese Philosophy*, Princeton University Press, Princeton.
Confucius (1971a). 'Confucian Analects,' in J. Legge (trans.), *Confucius: Confucian Analects, The Great Learning and The Doctrine of the Mean*, Dover, New York.
Confucius (1971b). 'Doctrine of the Mean,' in J. Legge (trans.), *Confucius: Confucian Analects, The Great Learning and The Doctrine of the Mean*, Dover, New York.
Confucius (1971c). 'The Great Learning in Confucius,' in J. Legge (trans.), *Confucius: Confucian Analects, The Great Learning and The Doctrine of the Mean*, Dover, New York.
Engelhardt, H.T., Jr. (2007). 'The Family in Transition and in Authority: The Impact of Biotechnology,' in S. Lee (Ed.), *The Family, Medical Decision-Making, and Biotechnology: Critical Reflections on Asian Moral Perspectives*, Springer, Dordrecht.
Fan, R. (1999). 'Just Health Care, The Good Life, and Confucianism,' in R. Fan (Ed.), *Confucian Bioethics* (pp. 257–284), Kluwer Academic Publishers, Boston.
Fan, R. (2007). 'Confucian Familism and its Bioethical Implications,' in S.C. Lee (Ed.), *The Family, Medical Decision-Making, and Biotechnology: Critical Reflections on Asian Moral Perspectives*, Springer, Dordrecht.
Hui, E. (2004). 'Personhood and Bioethics: Chinese Perspective,' in R. Qui (Ed.), *Asian Bioethics* (pp. 29–44), Kluwer Academic Publishers, Boston.
Hui, E. (1999). 'Personhood and Bioethics: Chinese Perspective,' in R. Fan (Ed.), *Confucian Bioethics* (pp. 127–163), Kluwer Academic Publishers, Boston.
Ivanhoe, P.J. (2000). *Confucian Moral Self Cultivation*, Hackett, Indianapolis.
Lo, P.C. (1999). 'Confucian Views on Suicide and Their Implications for Euthanasia,' in R. Fan (Ed.), *Confucian Bioethics* (pp. 257–284), Kluwer Academic Publishers, Boston.
Low, L. (1998). 'Health Care in the Context of Social Security in Singapore,' *SOJOURN: Journal of Social Issues in Southeast Asia*, 13, 139–165.

Massaro, T.A. & Wong, Y.N. (1995). 'Positive Experience with Medical Savings Accounts in Singapore,' *Health Affairs*, 14, 267–269.

Mencius (1970). *The Works of Mencius*, J. Legge (trans.), Dover Publications, New York.

Ministry of Health. 'Report of the Cost Review Committee-Response of the Ministry of Health' [On-line]. Available: http://gov.sg/moh/mohiss/review.html. Accessed April 28, 2006.

Ni, P. (1999). 'Confucian Virtues and Personal Health,' in R. Fan (Ed.), *Confucian Bioethics* (pp. 27–44), Kluwer Academic Publishers, Boston.

Schwartz, B.I. (2004). *The World of Thought in Ancient China*, Belknap Press, Boston.

Shun, K. (2002). 'Ren and Li in the Analects,' in B.W. Van Norden (Ed.), *Confucius and the Analects*, Oxford University Press.

Teo, K.S. (2007). 'Confucian Health Care System in Singapore: A Family-oriented Approach to Financial Sustainability' in S.C. Lee (Ed.), *The Family, Medical Decision-Making, and Biotechnology: Critical Reflections on Asian Moral Perspectives*, Springer, Dordrecht.

Van Norden, B.W. (2002). 'Introduction,' in B.W. Van Norden (Ed.), *Confucius and the Analects*, Oxford University Press, New York.

Waley, A.D. (1989). *The Analects of Confucius*, First Vintage Books, New York.

INDEX

217

Philosophy and Medicine

Philosophy and Medicine

21. G.J. Agich and C.E. Begley (eds.): *The Price of Health.* 1986
ISBN 90-277-2285-4
22. E.E. Shelp (ed.): *Sexuality and Medicine.* Vol. I: Conceptual Roots. 1987
ISBN 90-277-2290-0; Pb 90-277-2386-9
23. E.E. Shelp (ed.): *Sexuality and Medicine.* Vol. II: Ethical Viewpoints in Transition.
1987 ISBN 1-55608-013-1; Pb 1-55608-016-6
24. R.C. McMillan, H. Tristram Engelhardt, Jr., and S.F. Spicker (eds.): *Euthanasia and the Newborn.* Conflicts Regarding Saving Lives. 1987
ISBN 90-277-2299-4; Pb 1-55608-039-5
25. S.F. Spicker, S.R. Ingman and I.R. Lawson (eds.): *Ethical Dimensions of Geriatric Care.* Value Conflicts for the 21st Century. 1987 ISBN 1-55608-027-1
26. L. Nordenfelt: *On the Nature of Health.* An Action-Theoretic Approach. 2nd,
rev. ed. 1995 ISBN 0-7923-3369-1; Pb 0-7923-3470-1
27. S.F. Spicker, W.B. Bondeson and H. Tristram Engelhardt, Jr. (eds.): *The Contraceptive Ethos.* Reproductive Rights and Responsibilities. 1987
ISBN 1-55608-035-2
28. S.F. Spicker, I. Alon, A. de Vries and H. Tristram Engelhardt, Jr. (eds.): *The Use of Human Beings in Research.* With Special Reference to Clinical Trials. 1988
ISBN 1-55608-043-3
29. N.M.P. King, L.R. Churchill and A.W. Cross (eds.): *The Physician as Captain of the Ship.* A Critical Reappraisal. 1988 ISBN 1-55608-044-1
30. H.-M. Sass and R.U. Massey (eds.): *Health Care Systems.* Moral Conflicts in European and American Public Policy. 1988 ISBN 1-55608-045-X
31. R.M. Zaner (ed.): *Death: Beyond Whole-Brain Criteria.* 1988
ISBN 1-55608-053-0
32. B.A. Brody (ed.): *Moral Theory and Moral Judgments in Medical Ethics.* 1988
ISBN 1-55608-060-3
33. L.M. Kopelman and J.C. Moskop (eds.): *Children and Health Care.* Moral and Social Issues. 1989 ISBN 1-55608-078-6
34. E.D. Pellegrino, J.P. Langan and J. Collins Harvey (eds.): *Catholic Perspectives on Medical Morals.* Foundational Issues. 1989 ISBN 1-55608-083-2
35. B.A. Brody (ed.): *Suicide and Euthanasia.* Historical and Contemporary Themes.
1989 ISBN 0-7923-0106-4
36. H.A.M.J. ten Have, G.K. Kimsma and S.F. Spicker (eds.): *The Growth of Medical Knowledge.* 1990 ISBN 0-7923-0736-4
37. I. Löwy (ed.): *The Polish School of Philosophy of Medicine.* From Tytus Chałubiński (1820–1889) to Ludwik Fleck (1896–1961). 1990
ISBN 0-7923-0958-8
38. T.J. Bole III and W.B. Bondeson: *Rights to Health Care.* 1991
ISBN 0-7923-1137-X

Philosophy and Medicine

Philosophy and Medicine

Philosophy and Medicine

74. H.T. Engelhardt, Jr. and L.M. Rasmussen (eds.): *Bioethics and Moral Content: National Traditions of Health Care Morality*. Papers dedicated in tribute to Kazumasa Hoshino. 2002 [ASiB-3]   ISBN 1-4020-6828-2
75. L.S. Parker and R.A. Ankeny (eds.): *Mutating Concepts, Evolving Disciplines: Genetics, Medicine, and Society*. 2002   ISBN 1-4020-1040-0
76. W.B. Bondeson and J.W. Jones (eds.): *The Ethics of Managed Care: Professional Integrity and Patient Rights*. 2002   ISBN 1-4020-1045-1
77. K.L. Vaux, S. Vaux and M. Sternberg (eds.): *Convenants of Life. Contemporary Medical Ethics in Light of the Thought of Paul Ramsey*. 2002
                                                         ISBN 1-4020-1053-2
78. G. Khushf (ed.): *Handbook of Bioethics: Taking Stock of the Field from a Philosophical Perspective*. 2003   ISBN 1-4020-1870-3; Pb 1-4020-1893-2
79. A. Smith Iltis (ed.): *Institutional Integrity in Health Care*. 2003
                                                         ISBN 1-4020-1782-0
80. R.Z. Qiu (ed.): *Bioethics: Asian Perspectives A Quest for Moral Diversity*. 2003 [ASiB-4]   ISBN 1-4020-1795-2
81. M.A.G. Cutter: *Reframing Disease Contextually*. 2003   ISBN 1-4020-1796-0
82. J. Seifert: *The Philosophical Diseases of Medicine and Their Cure*. Philosophy and Ethics of Medicine, Vol. 1: Foundations. 2004   ISBN 1-4020-2870-9
83. W.E. Stempsey (ed.): *Elisha Bartlett's Philosophy of Medicine*. 2004 [CoME-2]
                                                         ISBN 1-4020-3041-X
84. C. Tollefsen (ed.): *John Paul II's Contribution to Catholic Bioethics*. 2005 [CSiB-3]   ISBN 1-4020-3129-7
85. C. Kaczor: *The Edge of Life*. Human Dignity and Contemporary Bioethics. 2005 [CSiB-4]   ISBN 1-4020-3155-6
86. R. Cooper: *Classifying Madness*. A Philosophical Examination of the Diagnostic and Statistical Manual of Mental Disorders. 2005   ISBN 1-4020-3344-3
87. L. Rasmussen (ed.): *Ethics Expertise*. History, Contemporary Perspectives, and Applications. 2005   ISBN 1-4020-3819-4
88. M.C. Rawlinson and S. Lundeen (eds.): *The Voice of Breast Cancer in Medicine and Bioethics*. 2006   ISBN 1-4020-4508-5
89. M. Bormuth (ed.): *Life Conduct in Modern Times: Karl Jaspers and Psychoanalysis*. 2006
                                                         ISBN 1-4020-4764-9
90. H. Kincaid and J. McKitrick (eds.): *Establishing Medical Reality: Essays in the Metaphysics and Epistemology of Biomedical Science*. 2006   ISBN 1-4020-5215-4
91. S.C. Lee (ed.): *The Family, Medical Decision-Making, and Biotechnology: Critical Reflections on Asian Moral Perspectives*. 2007   ISBN 978-1-4020-5219-4

Printed in the United States
87379LV00003B/344/A

9 781402 052194